Chinese Abacus For Kids

A step-by-step guide to addition and subtraction using the Chinese abacus (Suan Pan).

Author:
Paul Green

First published 2016

ISBN-13: 978-1540888693
ISBN-10: 154088869X

PREFACE

The Chinese abacus has been used as a calculation tool for generations and can still be seen in use in China today. Children are still taught to use this instrument in schools today. It is widely available, cheap to buy and fun to use.

This book will teach children the skills required to use the abacus effectively, once learnt and practised these skills will stay with them throughout their lives. A useful and impressive skill that would be an asset for anyone.

I'll show you.
Important
and give you tips.
I'll help you learn to use the Abacus.

CONTENTS

INTRODUCTION

Nice to know

- The Chinese abacus is also called the Suan Pan.
- This is abacus written in Chinese 算盤
- The Chinese abacus is mostly used for adding and subtracting numbers.
- The Chinese abacus has a wooden frame and seven beads per column, two beads above and five beads below.

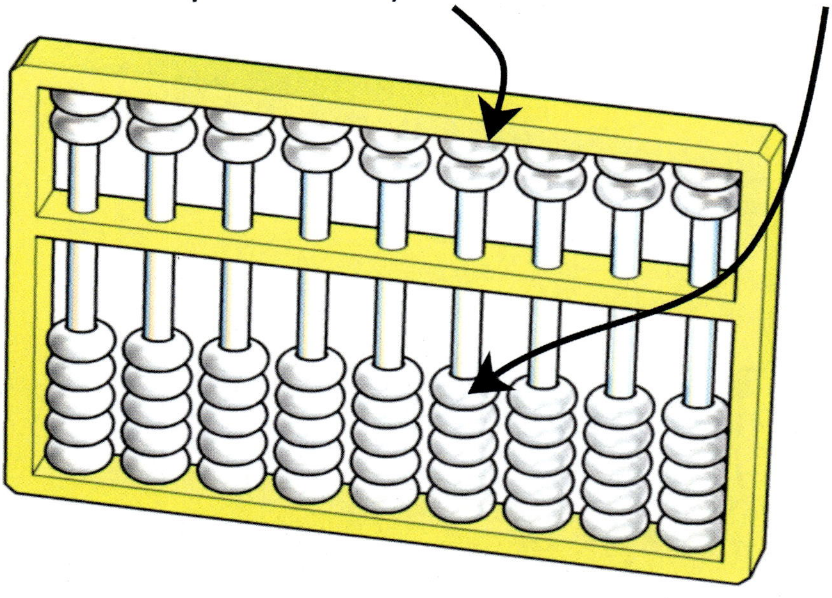

THE PARTS OF THE CHINESE ABACUS

1. A wooden frame.
2. A beam, to push the beads up against and away from.
3. Rods, to slide the beads up and down on.
4. 2 beads above the beam. Called 'Upper beads'.
5. 5 beads below the beam. Called 'Lower beads'.
6. A column is one rod and the 7 beads on that rod. There are 9 columns on this abacus. Abaci have different amounts of rods, usually 9 or 13 rods but some have less and some have more.
7. Upper base beads, all of the upper beads that touch the frame. No longer used, just leave these beads touching the frame. Shown in green below.
8. Lower base beads, all of the lower the beads that touch the frame. No longer used, just leave these beads touching the frame. Shown in red below.

Note Base beads are no longer used because we now use what is called the 'Decimal system', so we don't need them. The 'Decimal system' is the number system we use every day, based on 10 digits (0,1,2,3,4,5,6,7,8,9).

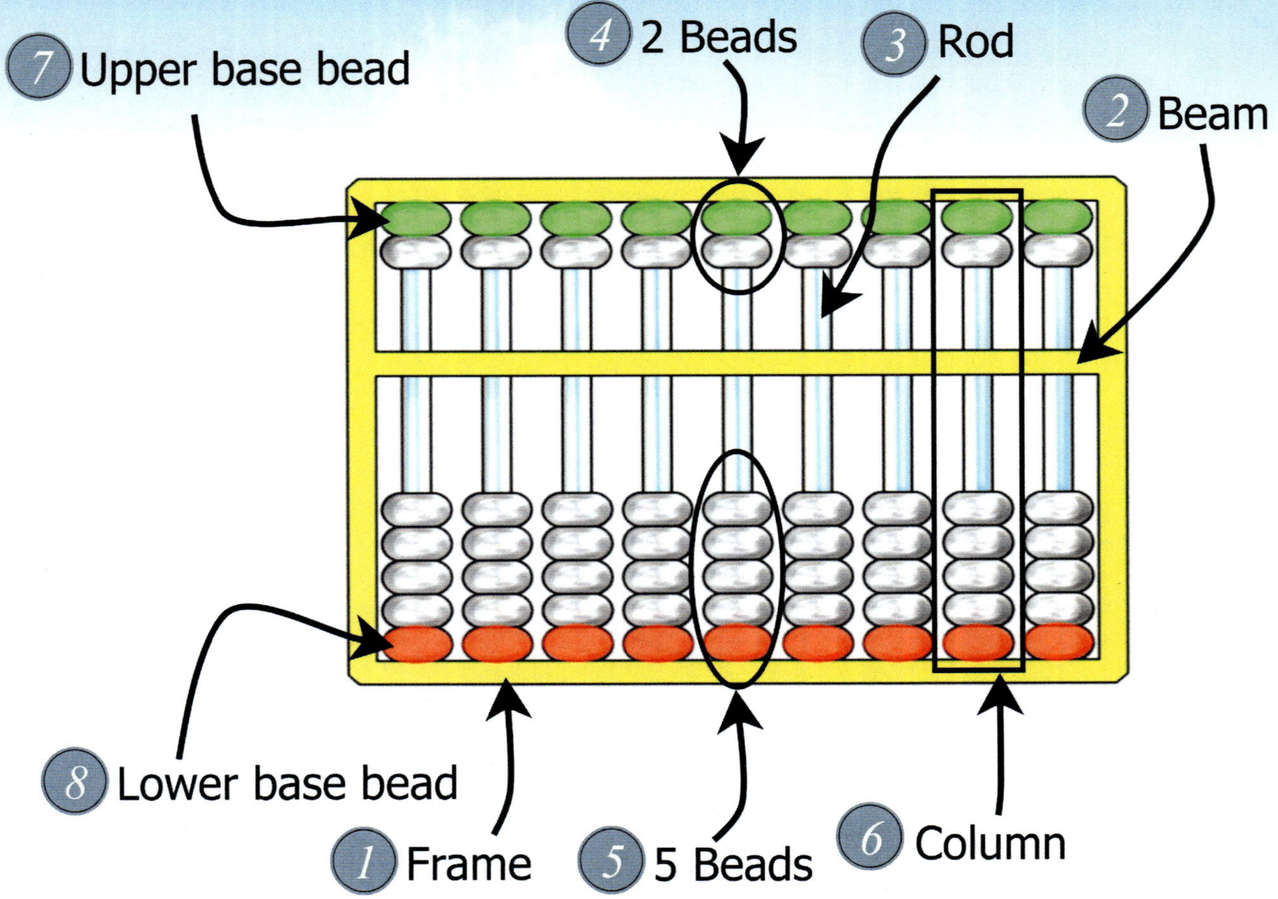

PUTTING YOUR NUMBERS IN THE CORRECT COLUMN

We need to know which abacus column to use to place each digit.

Let's look at the following 4 digit number **4213**

- The number 3 is on the 'Ones' column
- The number 1 is on the 'Tens' column
- The number 2 is on the 'Hundreds' column
- The number 4 is on the 'Thousands' column

4213

This is how it would look on the abacus.

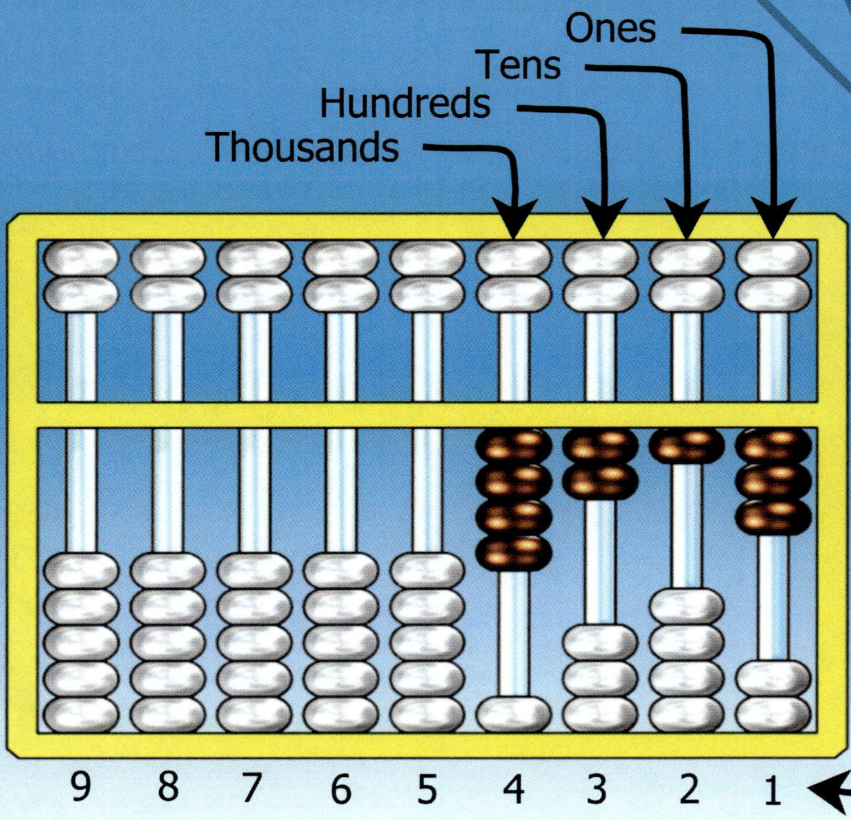

Tip Look how we start with the 'Ones' digit (3 in this example) on column 1.

WHAT AMOUNT IS EACH COLUMN WORTH?

The picture below shows the values of each column on the abacus.

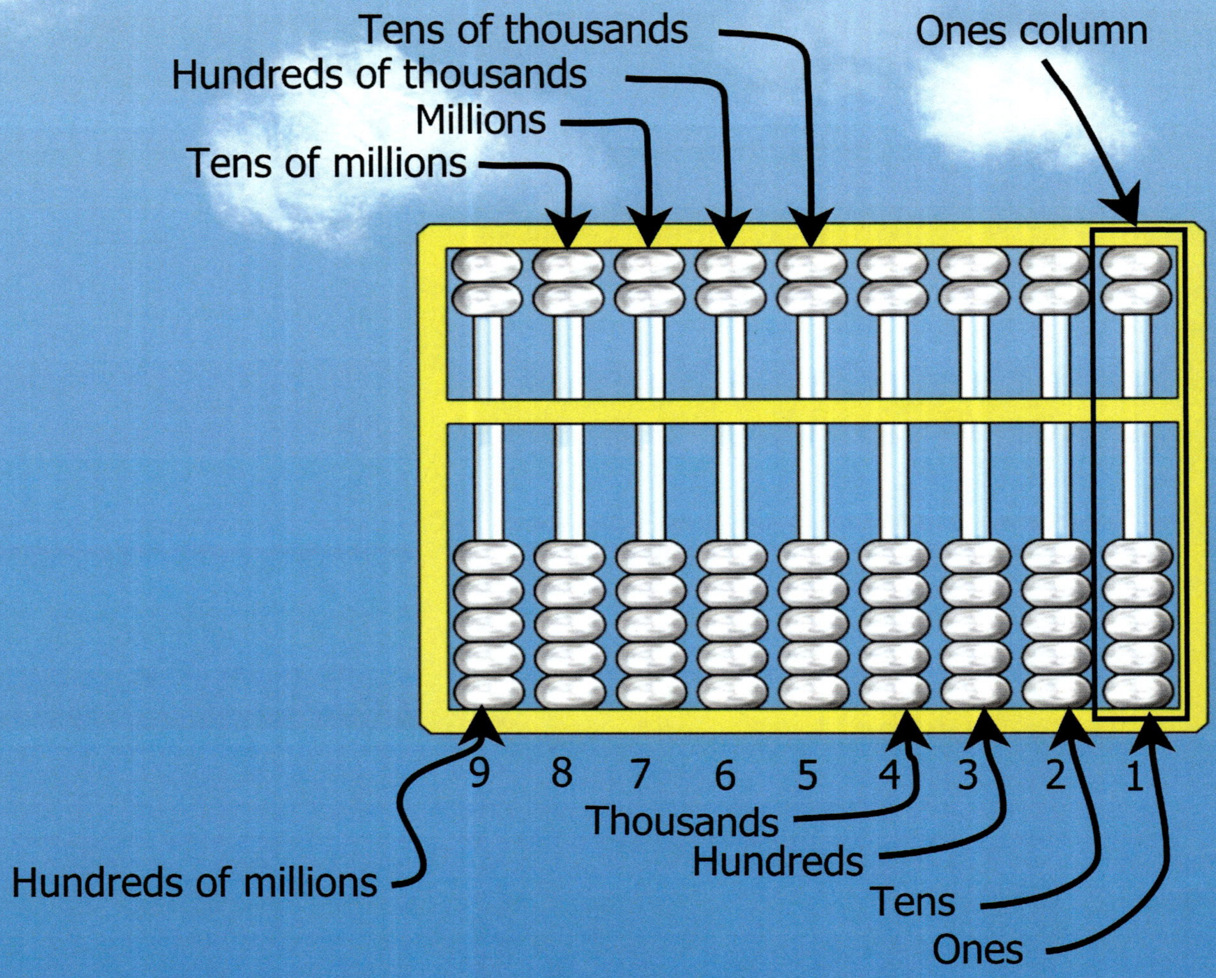

Tip Notice how each column value keeps getting TEN times BIGGER on the left of the ones column.

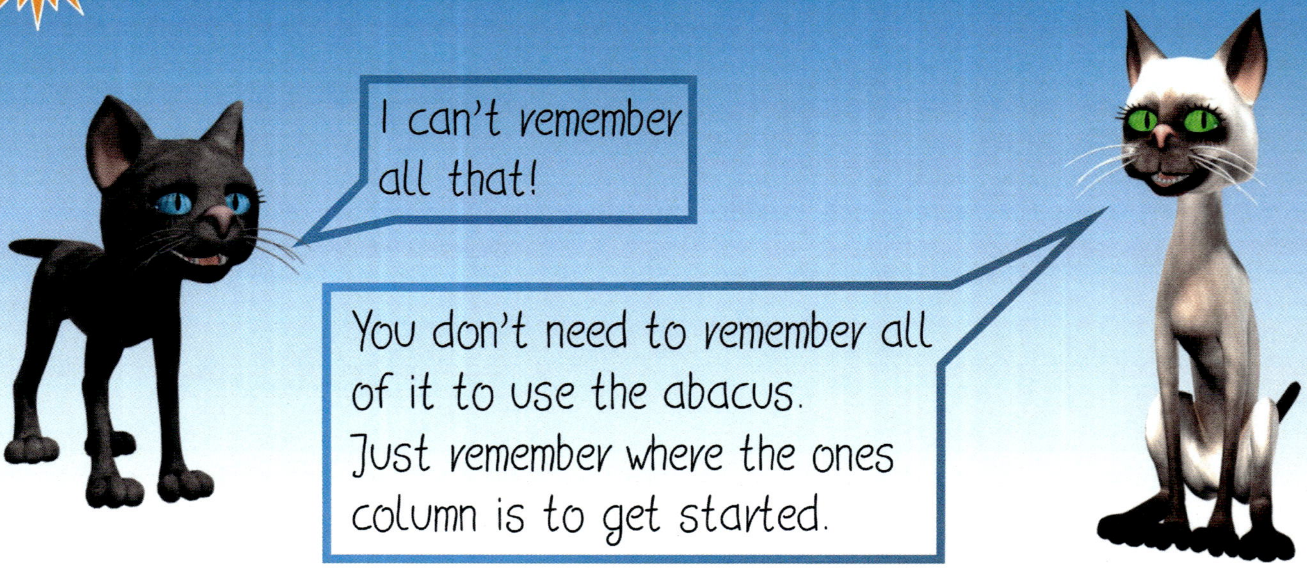

ABACUS BASICS

HOW TO MOVE BEADS TO MAKE A NUMBER ON THE ABACUS

Important!

- When we move a bead towards the beam this is called **'Register a bead'.**
- When we move a bead away from the beam this is called **'Unregister a bead'.**
- The bead above the beam is worth **5.**
- The beads below the beam are worth 1.
- We read the result on the abacus by looking only at the beads that are **pushed against the beam.**

Each columns result →

No beads registered (no beads pushed against the beam) →

Column number →

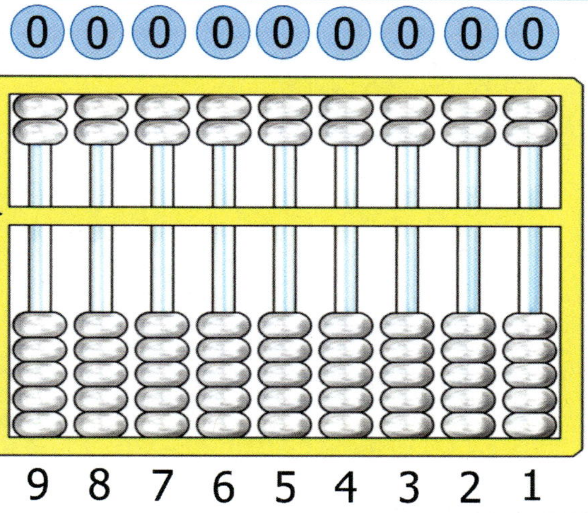

Tip

Always start with your abacus looking like this, reset all beads to *zero*

The abacus result is 0

Let's put the number 1 on the abacus

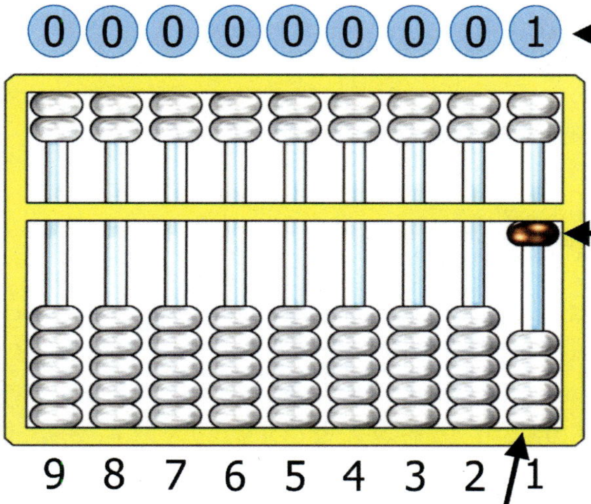

← Each columns result

← One bead registered (One lower bead is pushed against the beam)

Start at column number 1

Remember to start at the ones column, that's column 1, and on the lower deck.

The abacus result is 1

Let's put the number 5 on the abacus

Remember the upper bead is worth 5.

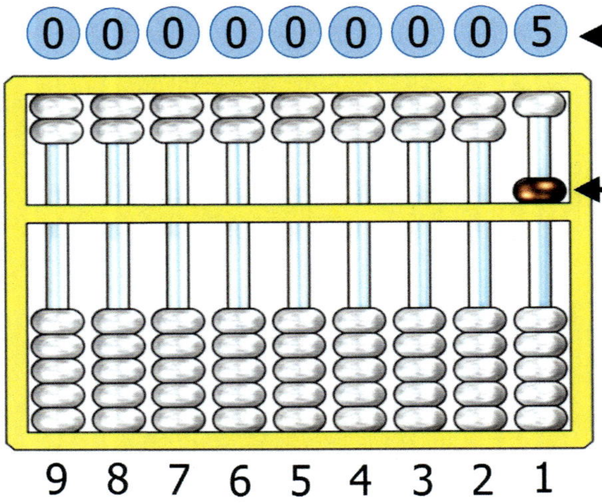

Each columns result

One bead registered (One upper bead is pushed against the beam)

The abacus result is 5

Let's put the number 6 on the abacus

The upper bead is worth 5, the lower bead is worth 1.

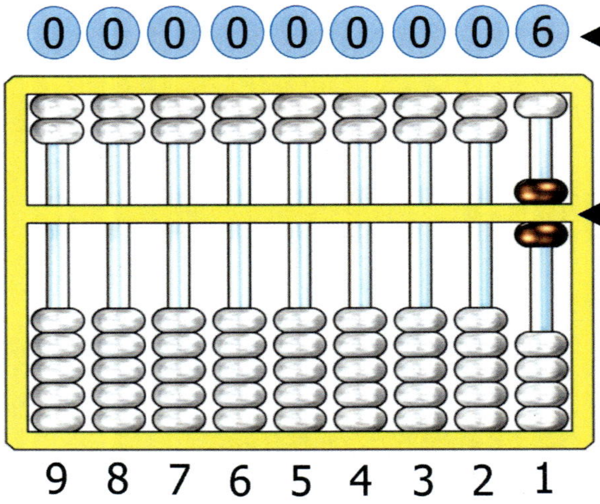

Each columns result

Two beads registered (One upper bead and one lower bead are pushed against the beam)

The abacus result is 6

Let's put the number 8 on the abacus

The upper bead is worth 5 plus 3 lower beads worth 1 each. This gives a total of 8.

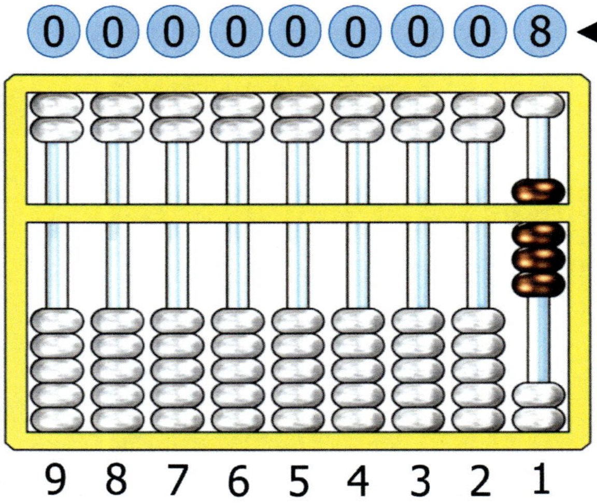

Each columns result

Four beads registered (One upper bead and three lower beads are pushed against the beam)

The abacus result is 8

Here are the single digit numbers on the abacus

0

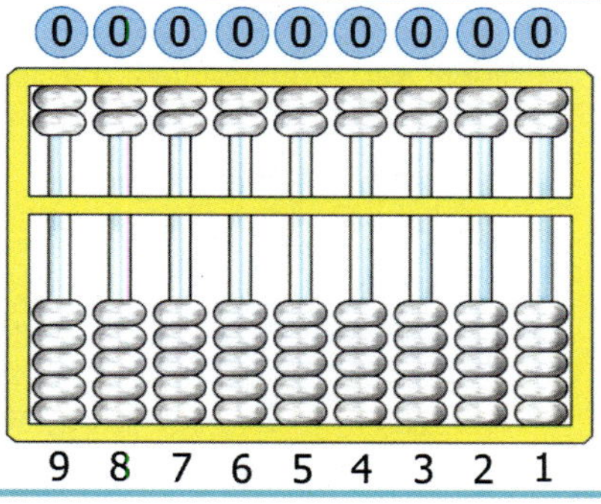

1

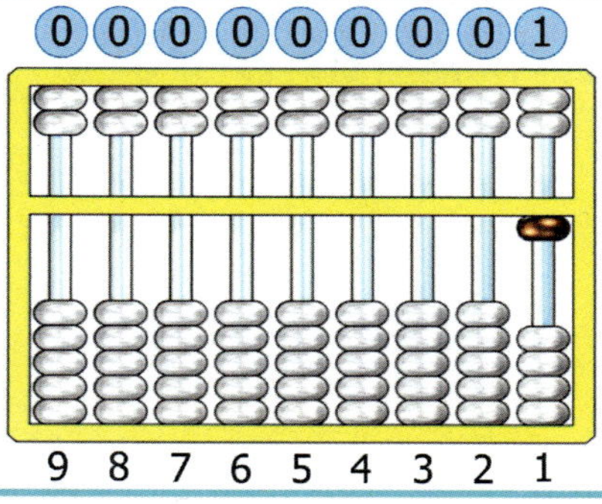

2

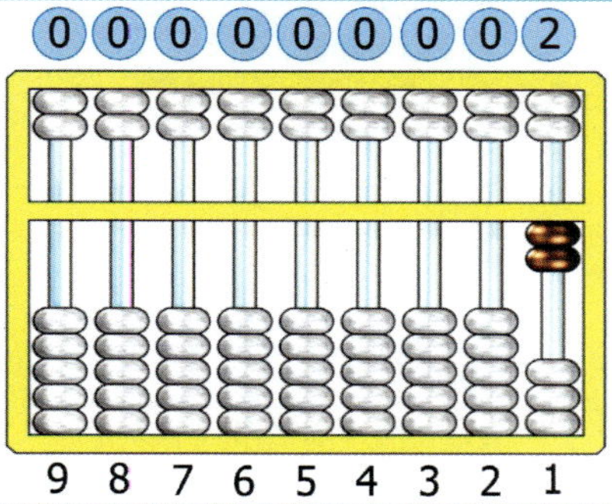

3

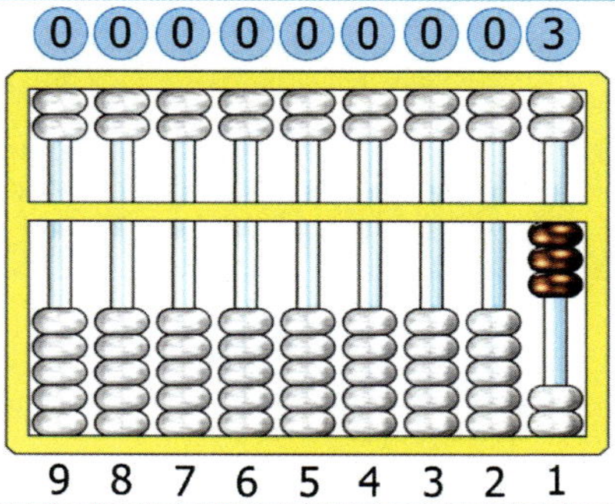

4

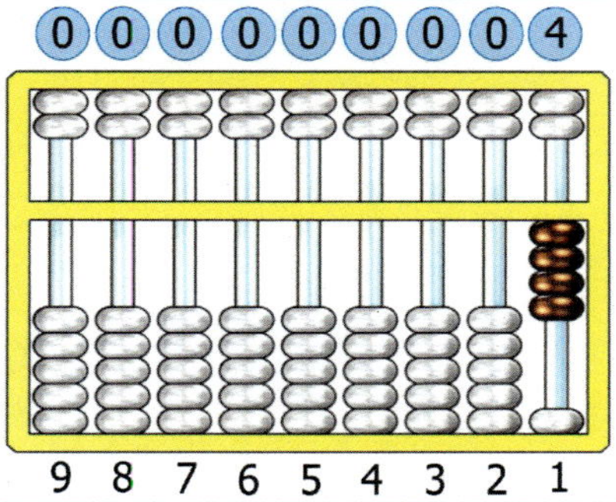

5

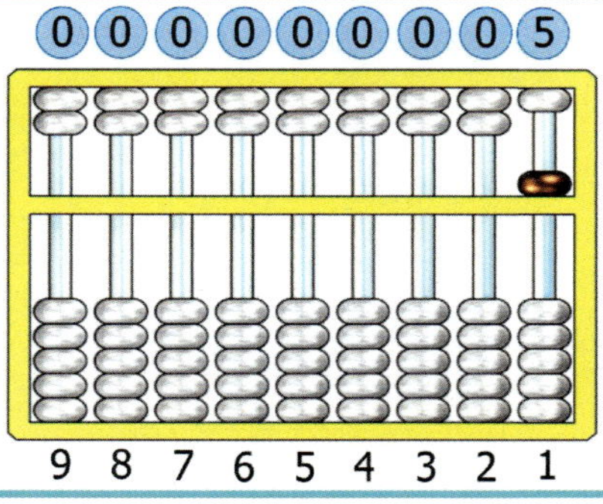

6

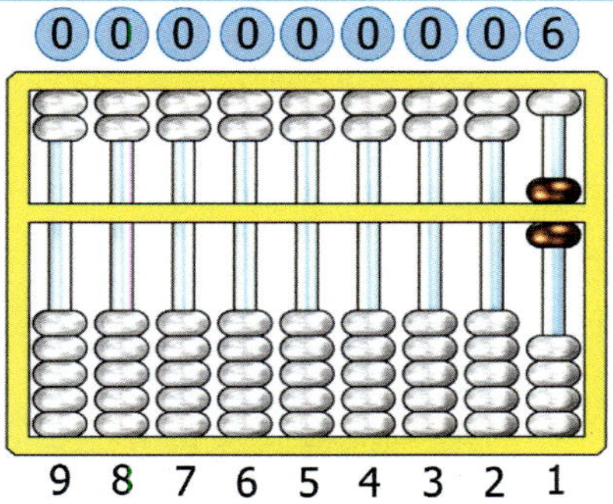

7

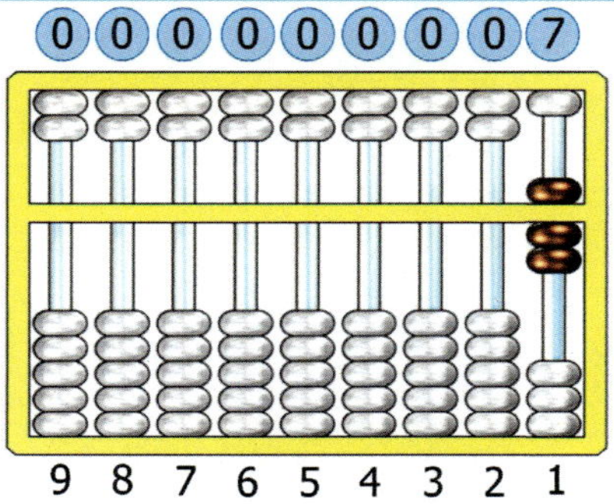

8

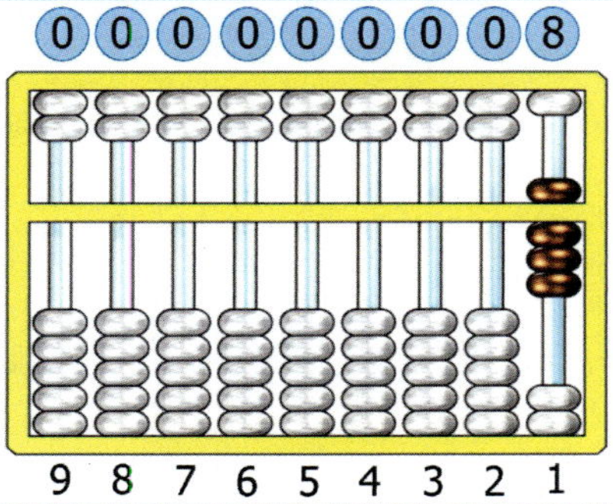

9

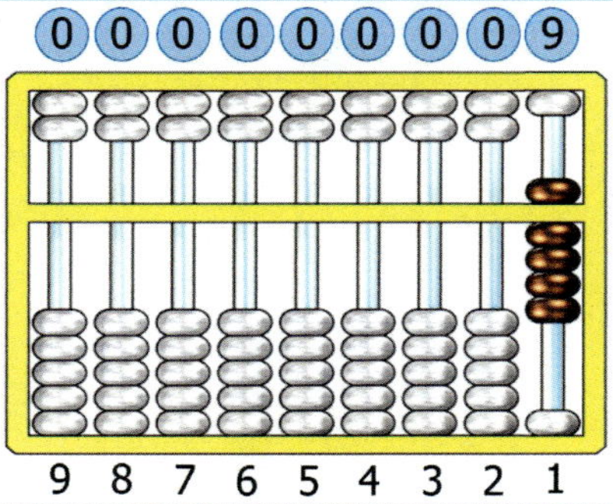

HOW TO PUT A NUMBER ON THE ABACUS WITH MORE THAN ONE DIGIT

Important!

- A **digit** is a symbol used to show a number. Example, **6** is one digit and is made up of one number.
- A digit is any number from **0** to **9**.
- The number **78** has two digits, **7** and **8**. There are two digits that make up the number 78.
- There are two places in the number 78, the ones place holds the number 8 and the tens place holds the number 7.

Let's put the number 15 on the abacus

When we put a number on the abacus or '**register**' a number by pushing the beads towards the beam, we start with the leftmost digit first (in this example the 1 of the 15) then move to the right to register the other numbers.

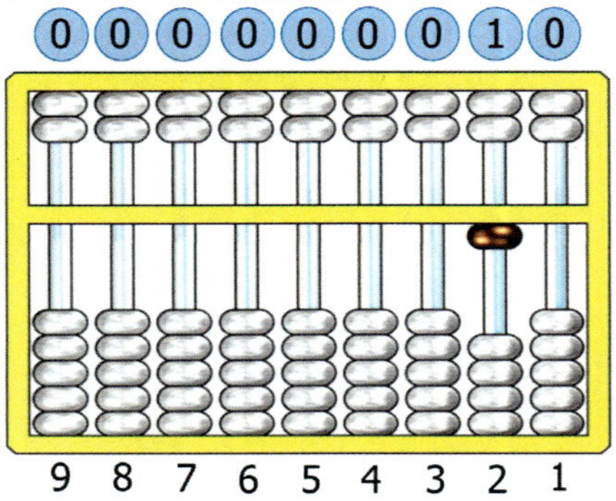

15

First register the '1' digit (leftmost) in the 'tens column', column 2.

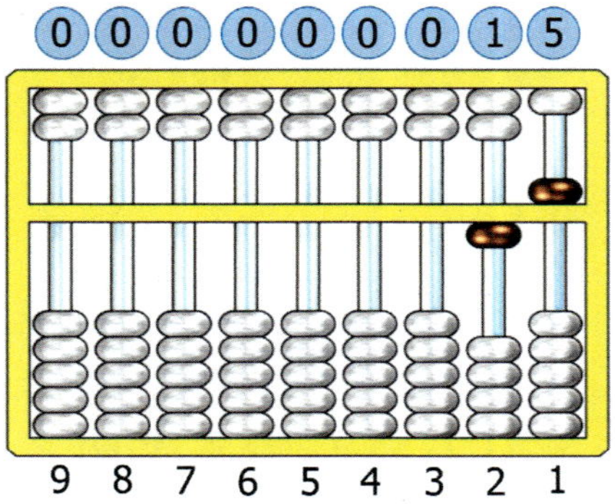

15

Second register the '5' digit in the 'ones column', column 1.

The abacus result is 15

Remember, we start with the leftmost digit first!
The number 1 of the 15 is in the tens column,
the number 5 of the 15 is in the ones column.

Here are some two digit numbers on the abacus

45

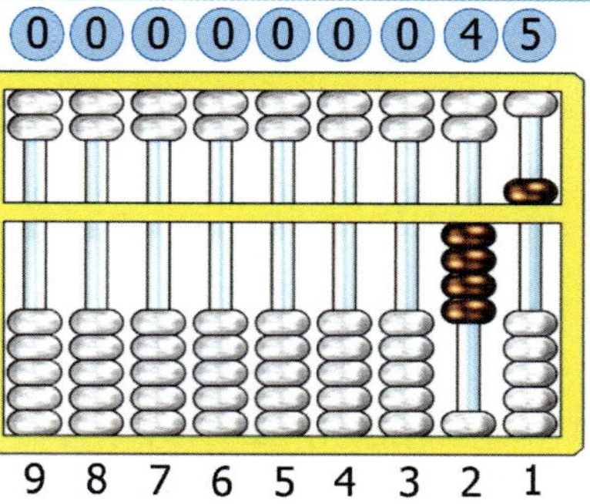

- We will put 45 on the abacus
- 45 has 2 digits, so use 2 columns
- Column 2, register 4 lower beads (this is for the 4 of the 45)
- Column 1, register 1 upper bead (this is for the 5 of the 45)

The abacus result is 45

14

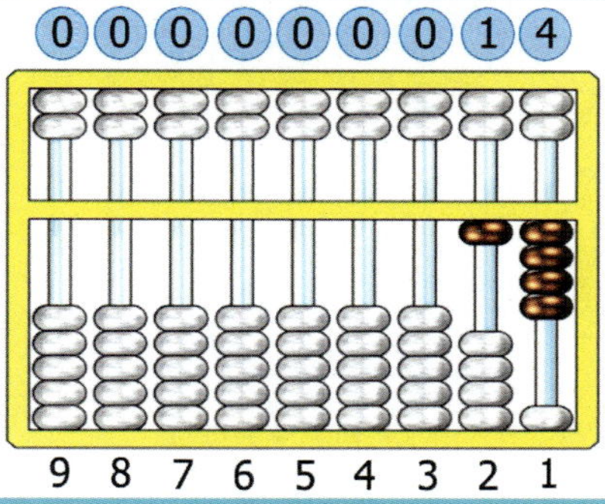

- We will put 14 on the abacus
- 14 has 2 digits, so use 2 columns
- Column 2, register 1 lower bead (this is for the 1 of the 14)
- Column 1, register 4 lower beads (this is for the 4 of the 14)

The abacus result is 14

95

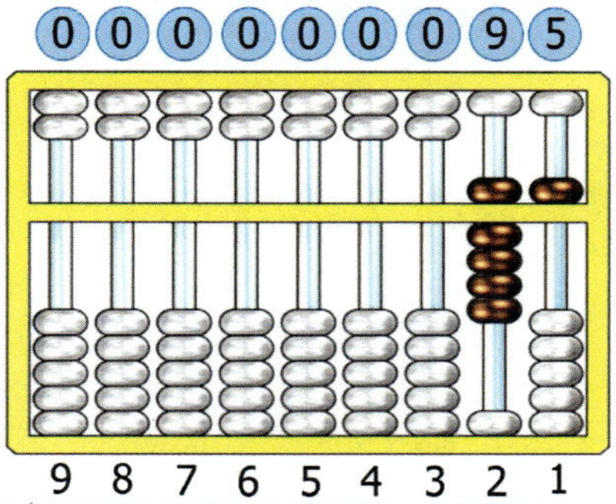

- We will put 95 on the abacus
- 95 has 2 digits, so use 2 columns
- Column 2, register 1 upper bead and 4 lower beads (this is for the 9 of the 95) Total on this column is 5+4=9
- Column 1, register 1 upper bead (this is for the 5 of the 95)

The abacus result is 95

56

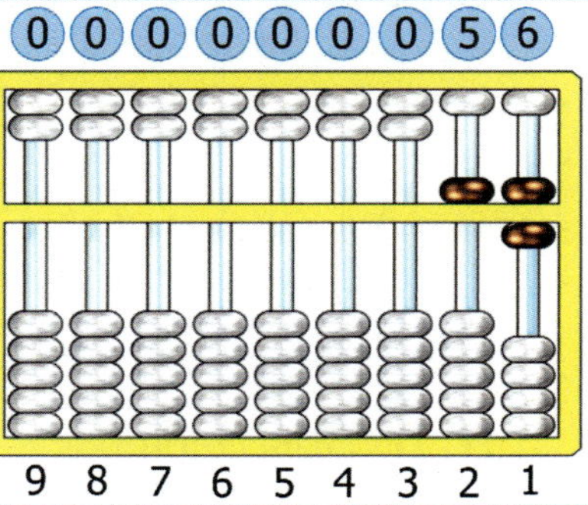

- We will put 56 on the abacus
- 56 has 2 digits, so use 2 columns
- Column 2, register 1 upper bead (this is for the 5 of the 56)
- Column 1, register 1 upper bead and 1 lower bead (this is for the 6 of the 56) Total on this column is 5+1=6

The abacus result is 56

87

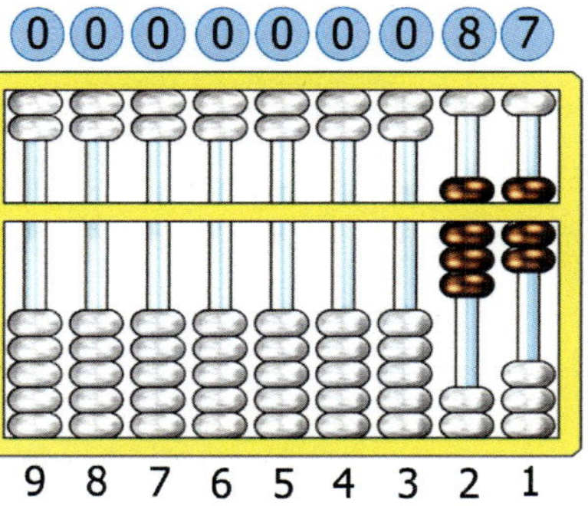

- We will put 87 on the abacus
- 87 has 2 digits, so use 2 columns
- Column 2, register 1 upper bead and 3 lower beads (this is for the 8 of the 87). Total on this column is 5+3=8
- Column 1, register 1 upper bead and 2 lower beads (this is for the 7 of the 87). Total on this column is 5+2=7

The abacus result is 87

How to register multi-digit numbers on the abacus

A multi-digit number is any number that has more than one digit.

We have already looked at some multi-digit numbers on the previous page (two digit numbers).

Now we will look at larger numbers. Let's start with a 5 digit number.

23456

First register the '2' digit (leftmost) in the 5th column.

Why the 5th column? Because the number has 5 digits.

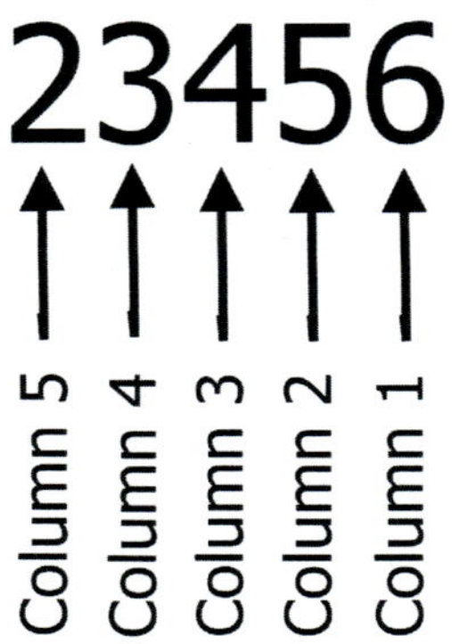

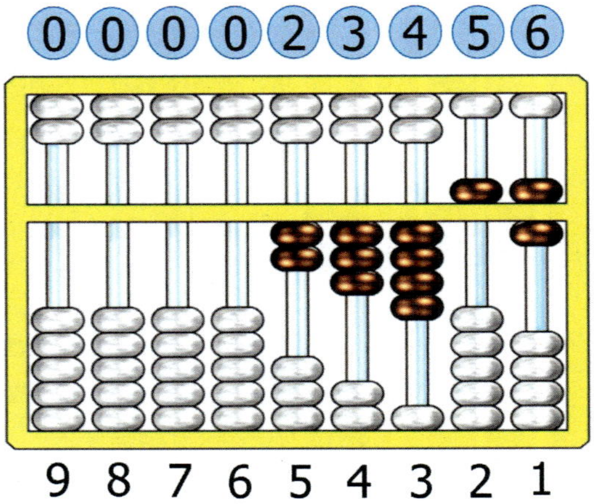

23456

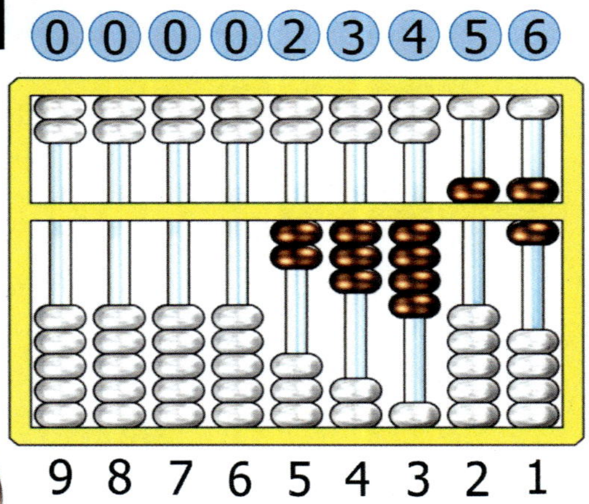

- We will put 23456 on the abacus
- 23456 has 5 digits, so use 5 columns (start on column 5)

- Column 5, register 2 lower beads
- Column 4, register 3 lower beads
- Column 3, register 4 lower beads
- Column 2, register 1 upper bead
- Column 1, register 1 upper bead and 1 lower bead
  (total on this column is 5+1=6)

The abacus result is 23456

Things to remember before we move on:

- Put all digits in their correct columns
- Register the leftmost digit first, then the next digit to the right and so on

Here are some multi-digit numbers on the abacus

556677

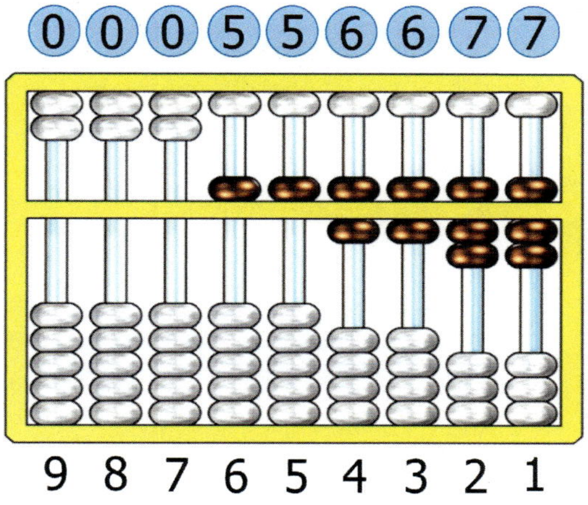

- We will put 556677 on the abacus
- 556677 has 6 digits, so use 6 columns (start on column 6)
- Column 6, register 1 upper bead
- Column 5, register 1 upper bead
- Column 4, register 1 upper bead and 1 lower bead
- Column 3, register 1 upper bead and 1 lower bead
- Column 2, register 1 upper bead and 2 lower beads
- Column 1, register 1 upper bead and 2 lower beads

The abacus result is 556677

316

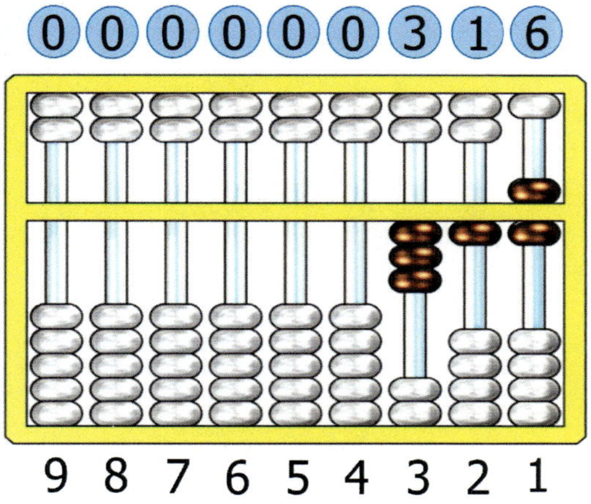

- We will put 316 on the abacus
- 316 has 3 digits, so use 3 columns (start on column 3)
- Column 3, register 3 lower beads
- Column 2, register 1 lower bead
- Column 1, register 1 upper bead and 1 lower bead

The abacus result is 316

289370078

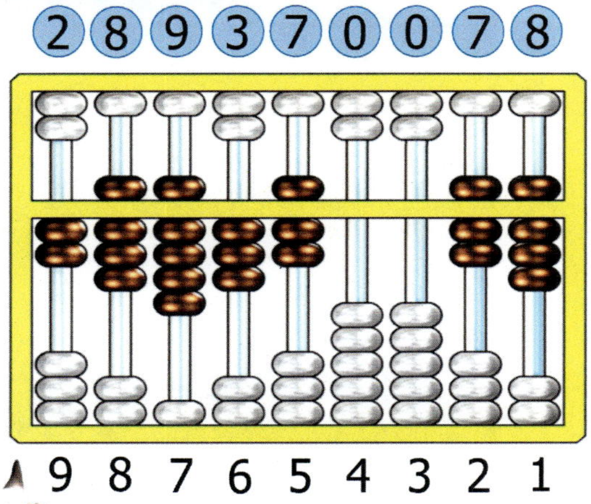

- We will put 289370078 on the abacus
- 289370078 has 9 digits, so use 9 columns (start on column 9)
- Column 9, register 2 lower beads
- Column 8, register 1 upper bead and 3 lower beads
- Column 7, register 1 upper bead and 4 lower beads
- Column 6, register 3 lower beads
- Column 5, register 1 upper bead and 2 lower beads
- Column 4, do nothing
- Column 3, do nothing
- Column 2, register 1 upper bead and 2 lower beads
- Column 1, register 1 upper bead and 3 lower beads

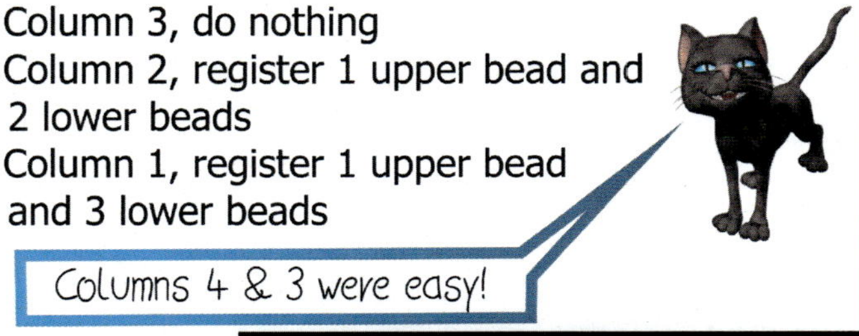

The abacus result is 289370078

TEST 1 - What numbers are shown on the abacus?

Answers are on page 20.

1

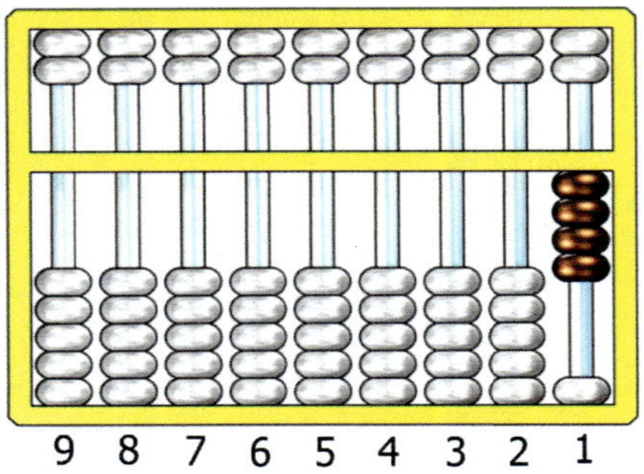

What number is on the abacus?

2

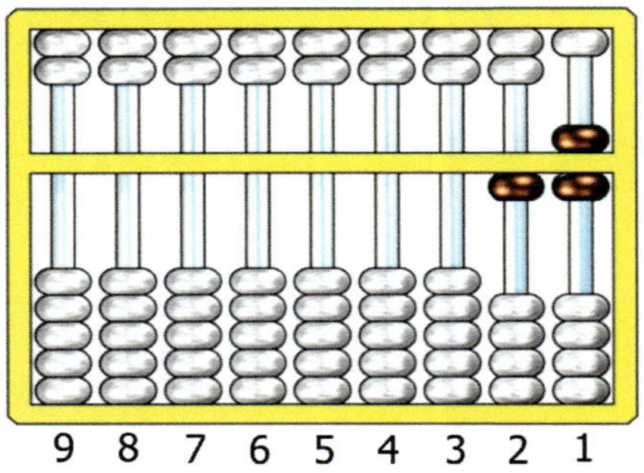

What number is on the abacus?

3

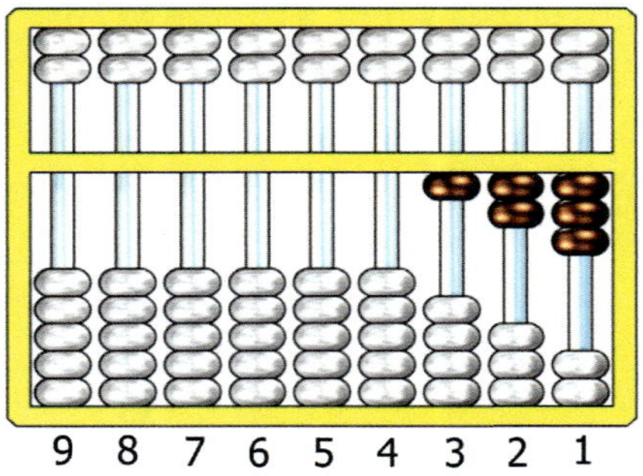

What number is on the abacus?

4

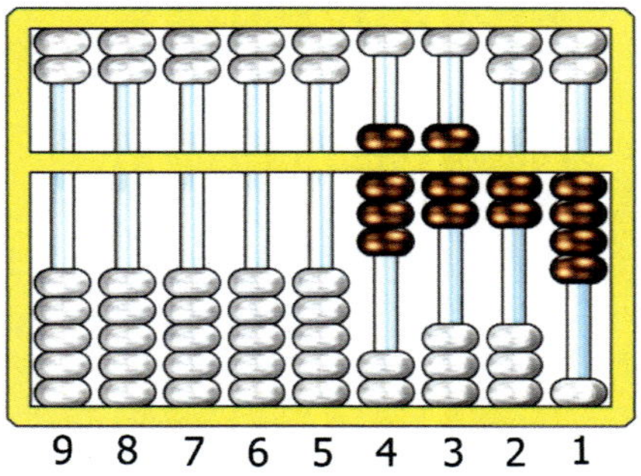

What number is on the abacus?

TEST 1 - What numbers are shown on the abacus?

Answers are on page 20.

5

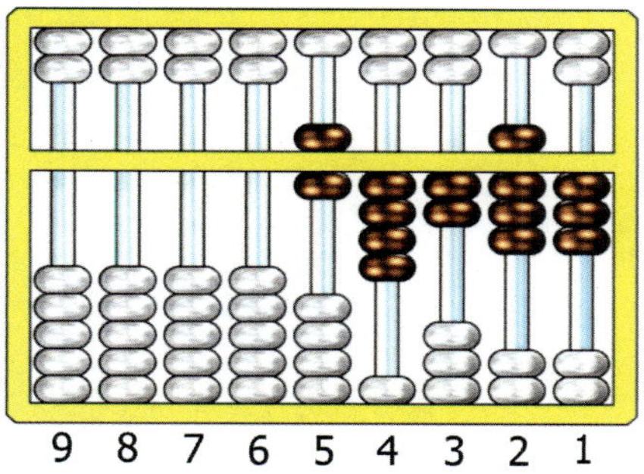

What number is on the abacus?

6

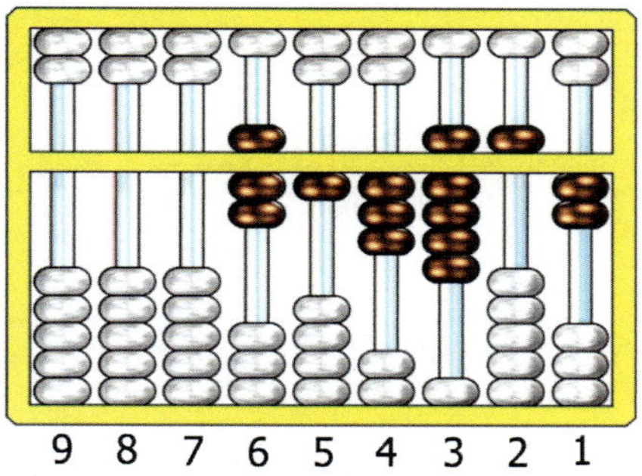

What number is on the abacus?

7

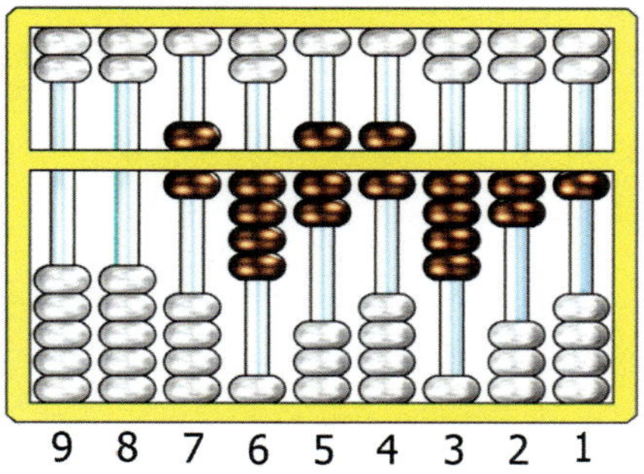

What number is on the abacus?

8

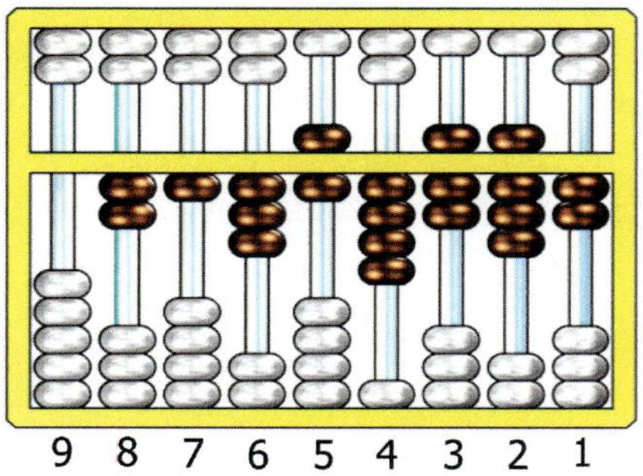

What number is on the abacus?

9

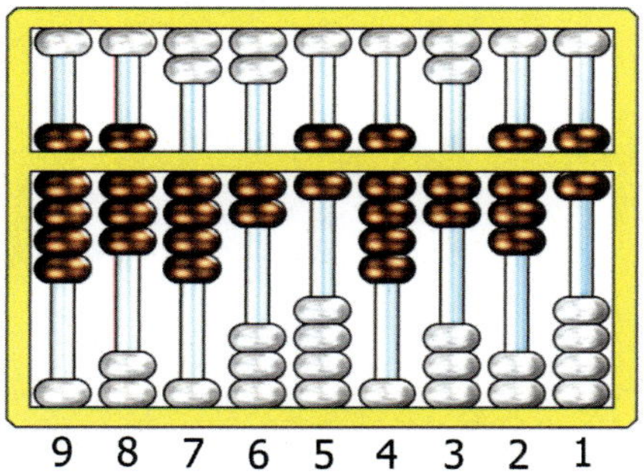

What number is on the abacus?

Answers to test 1 (on pages 18 & 19)

1	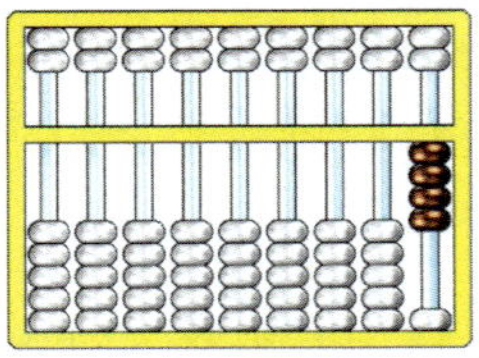	4
2	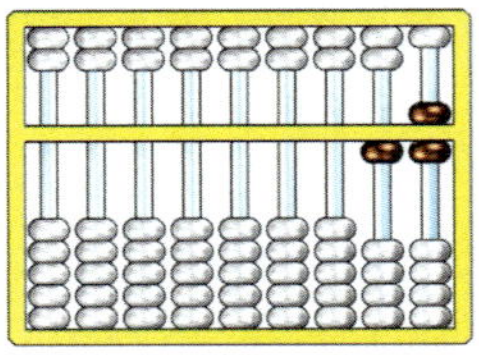	16
3	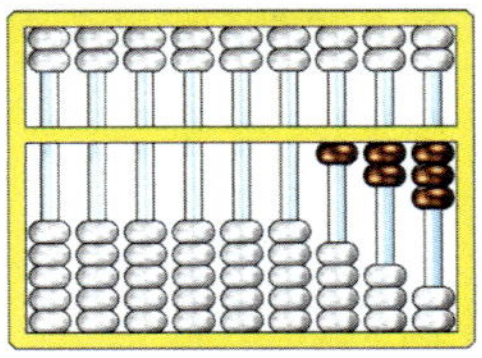	123
4	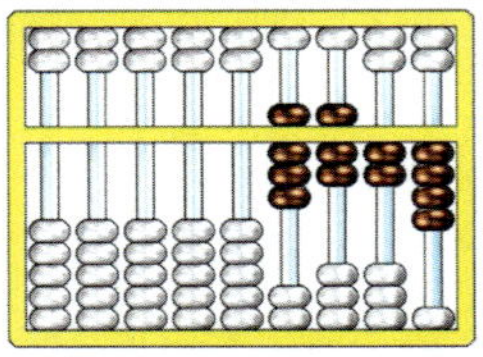	8724
5	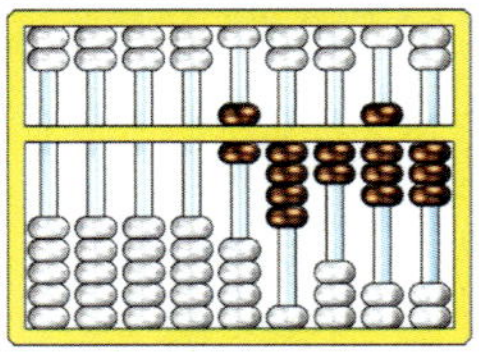	64283
6	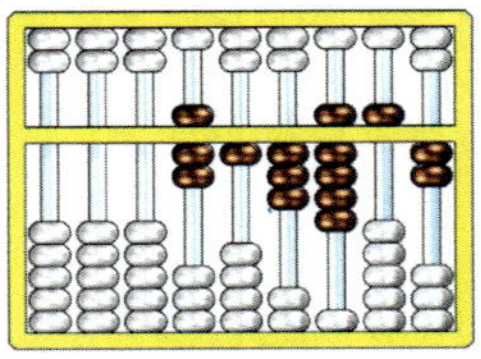	713952
7	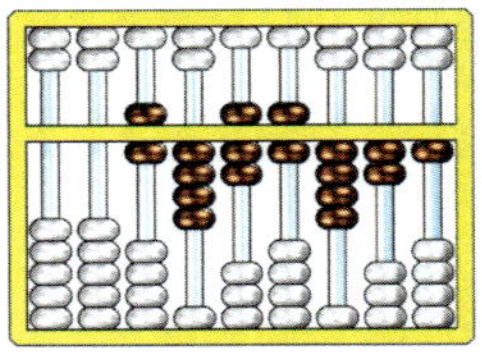	6476421
8	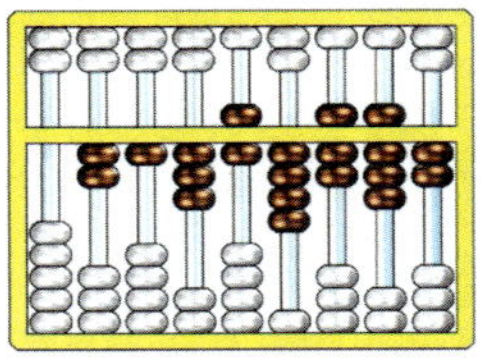	21364782
9	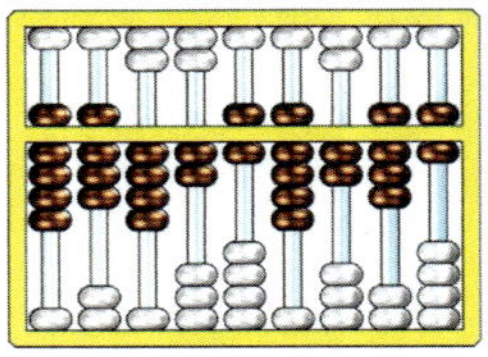	984269286

MOVING THE BEADS

What fingers do you use to move the beads?

The pictures on pages 21 to 25 will show you what fingers you use to register and unregister the beads.

We will use the thumb, index and middle finger. Some people like to use only the thumb and index finger, but we will use the middle finger also.

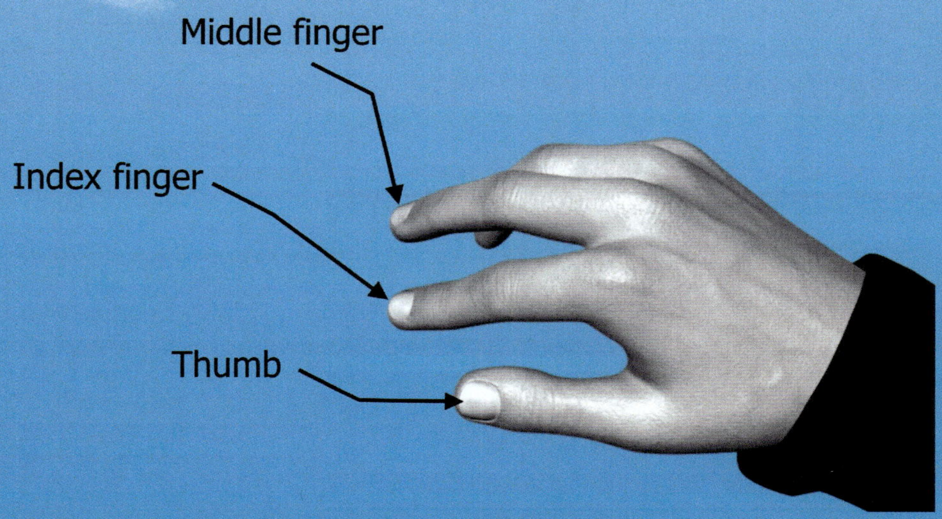

- **Thumb**
  Used to register the **LOWER** beads (to push them towards the beam).

- **Index finger**
  Used to unregister the **LOWER** beads (to move them away from the beam).

- **Middle finger**
  Used to register and unregister the **UPPER** beads (to move them to and away from the beam).

- **Bead order**
  When registering and unregistering a number, always start with the highest value column first, then work towards the lowest value column.
  For example, when registering the number 123 start by registering the 1 (100) then the 2 (20) and finally the 3 (3).

On the next 4 pages there are pictures which show the finger movements for registering and unregistering.

Register number 1

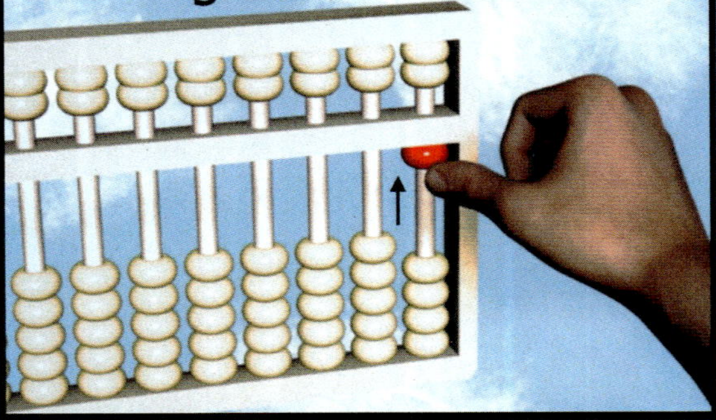

Register number 2

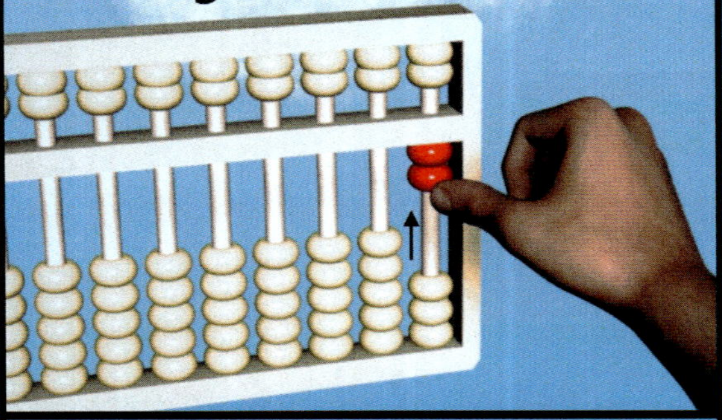

Register number 3

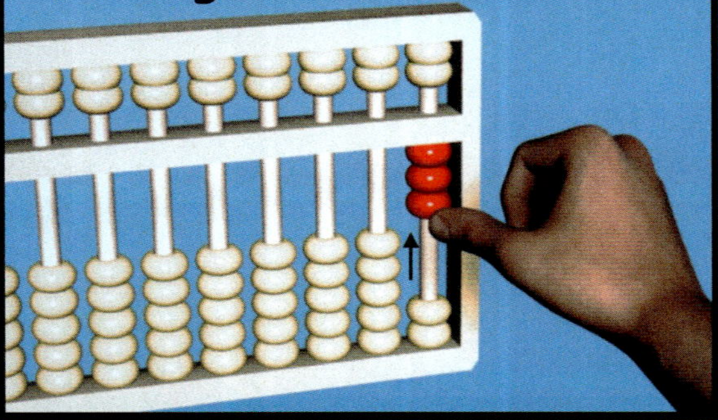

Register number 4

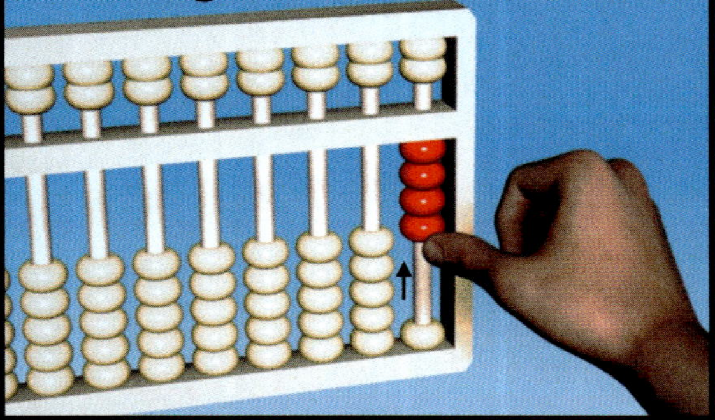

Register number 5

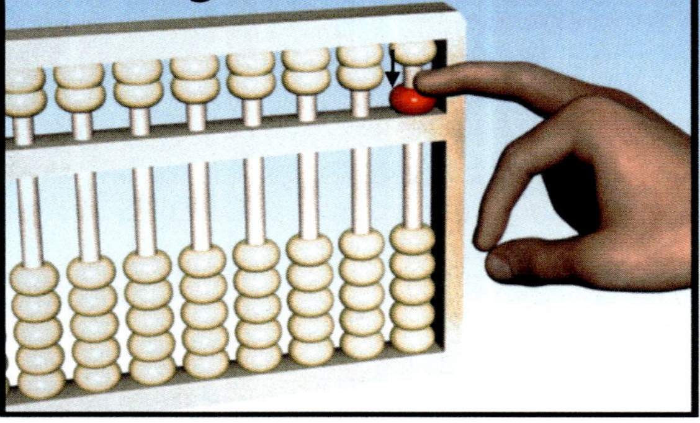

Unregister 1 from number 4

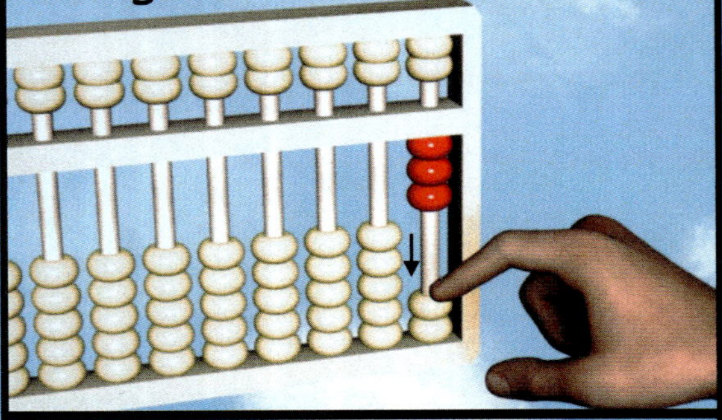

Unregister 2 from number 4

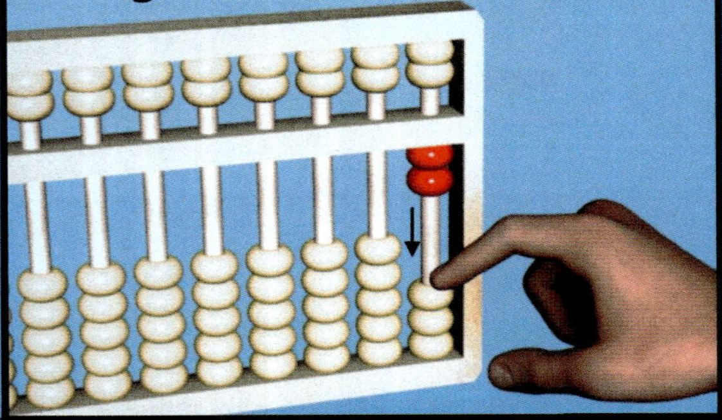

Unregister 3 from number 4

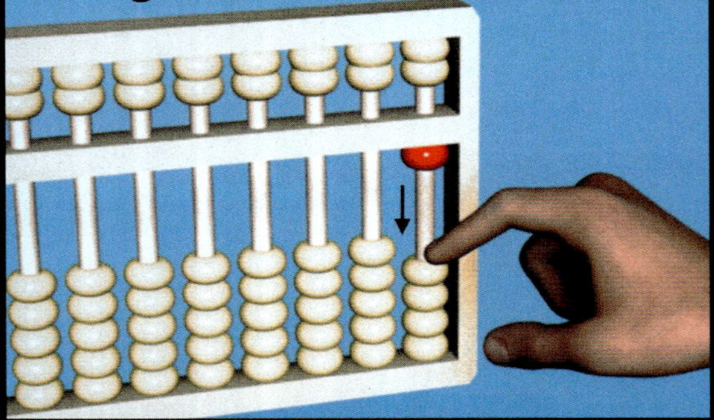

Unregister 4 from number 4

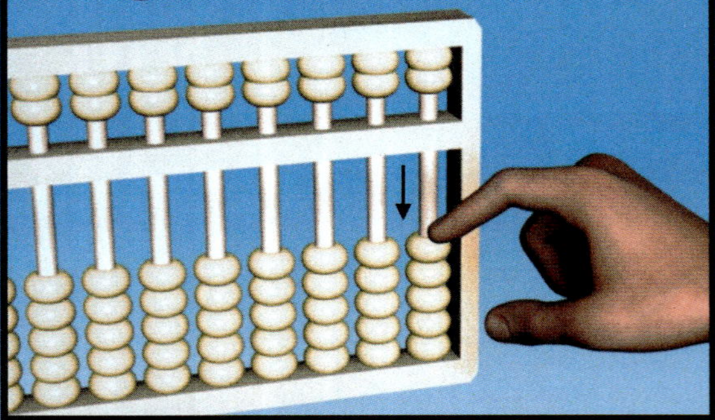

Unregister 5 from number 5

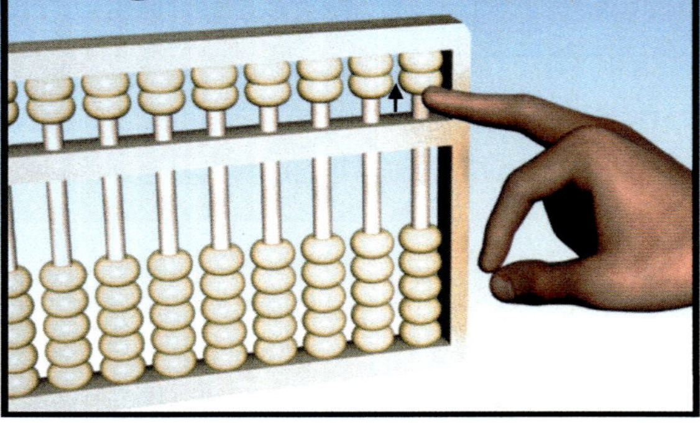

Register number 6

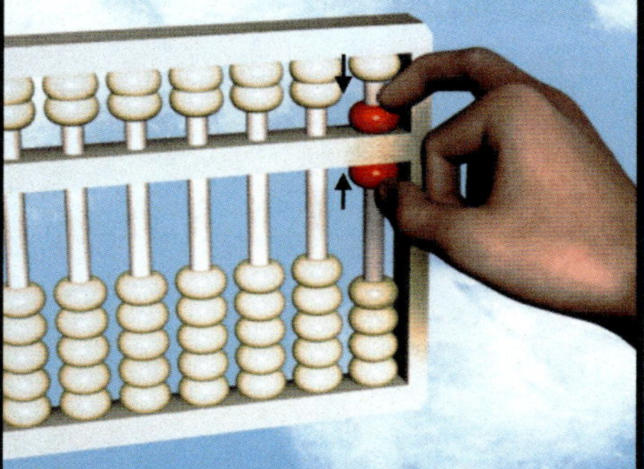

Pinch together in one move

Unregister 6 from number 6

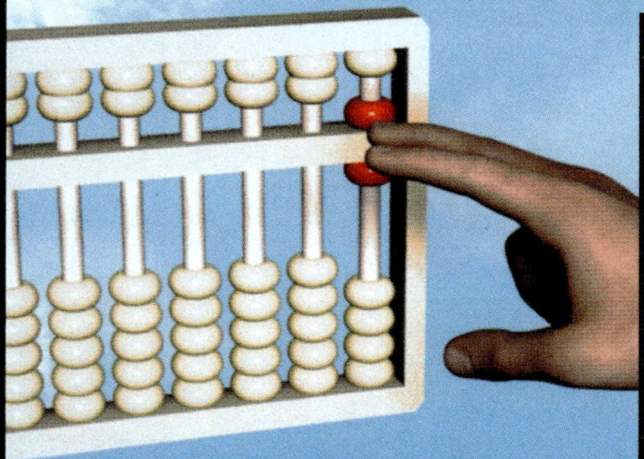

Fingers start position

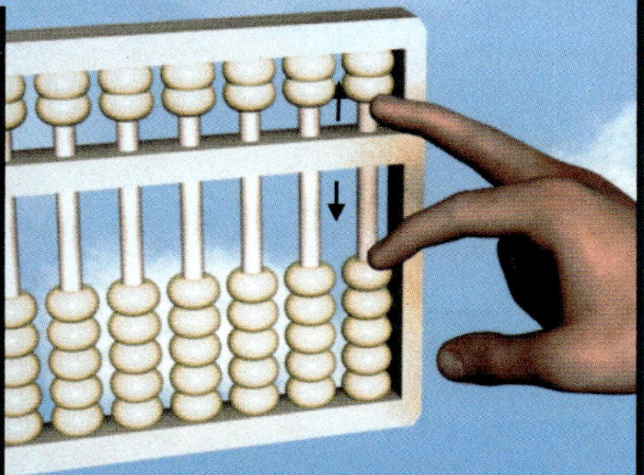

Separate in one move

Register number 7

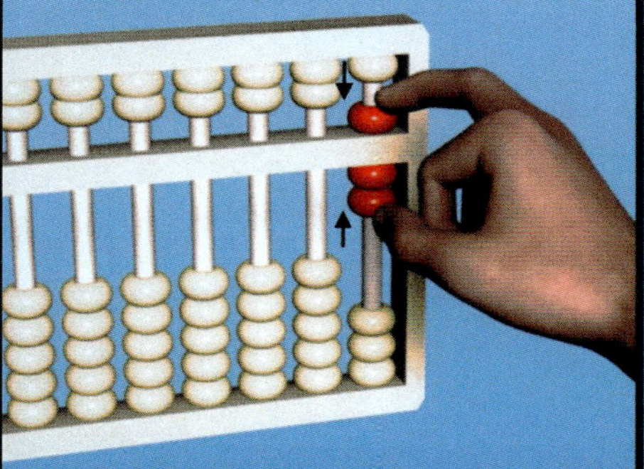

Pinch together in one move

Unregister 7 from number 7

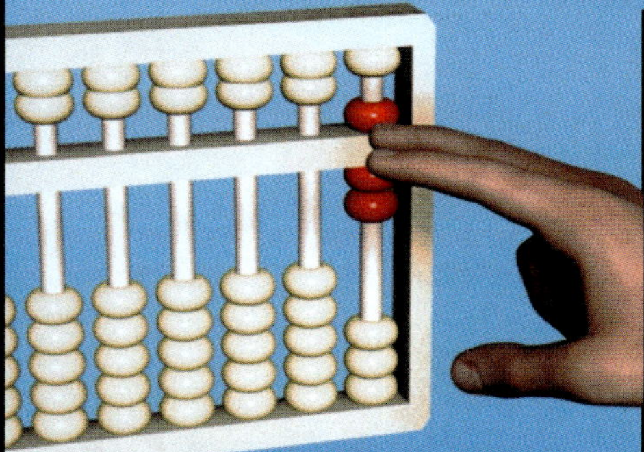

Fingers start position

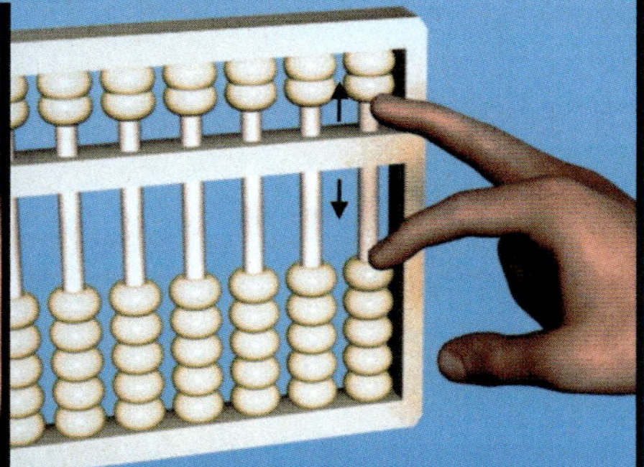

Separate in one move

Register number 8

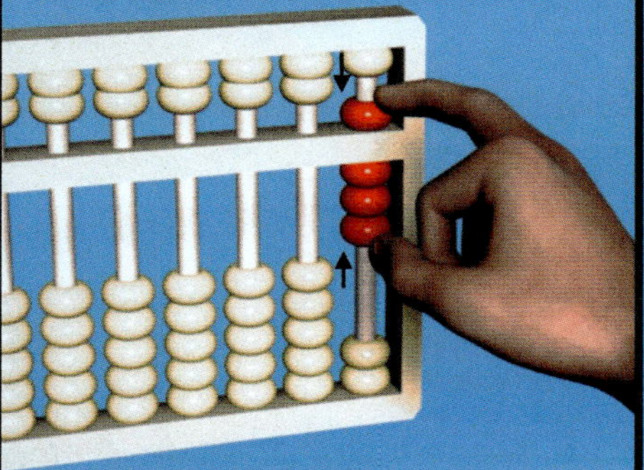

Pinch together in one move

Unregister 8 from number 8

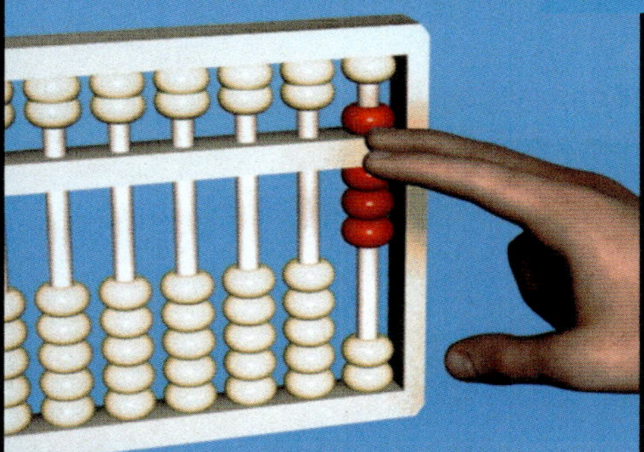

Fingers start position

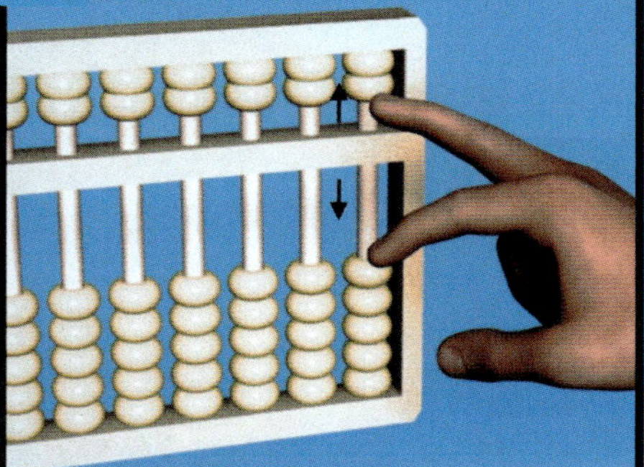

Separate in one move

Register number 9

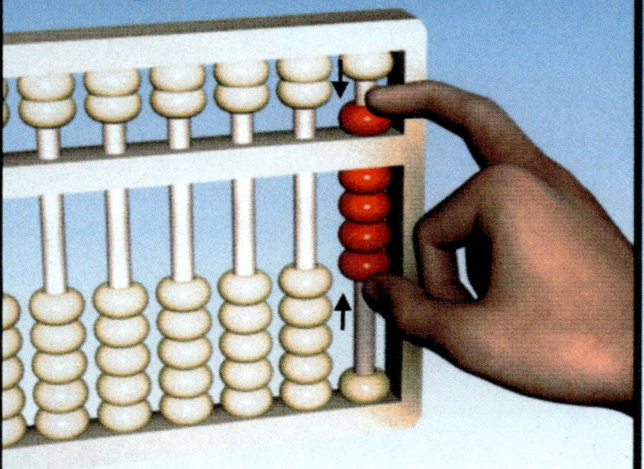

Pinch together in one move

Unregister 9 from number 9

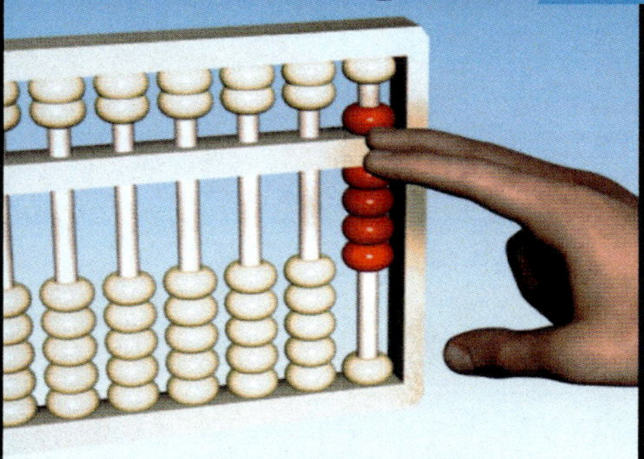

Fingers start position

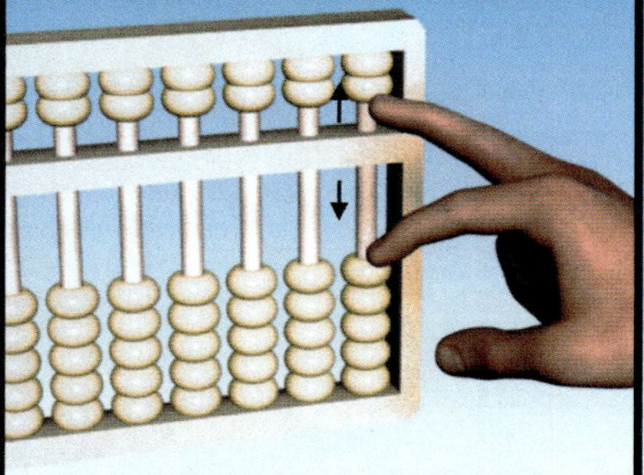

Separate in one move

Register number 10

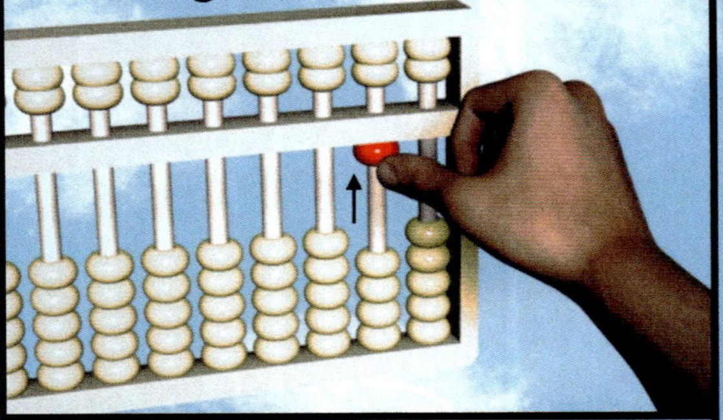

Register number 20

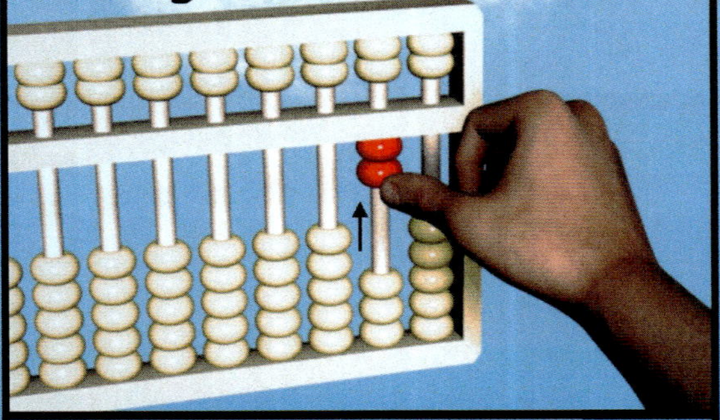

Register number 30

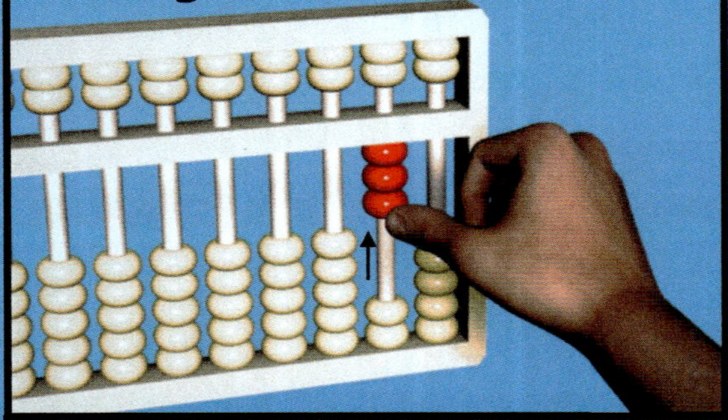

Register number 40

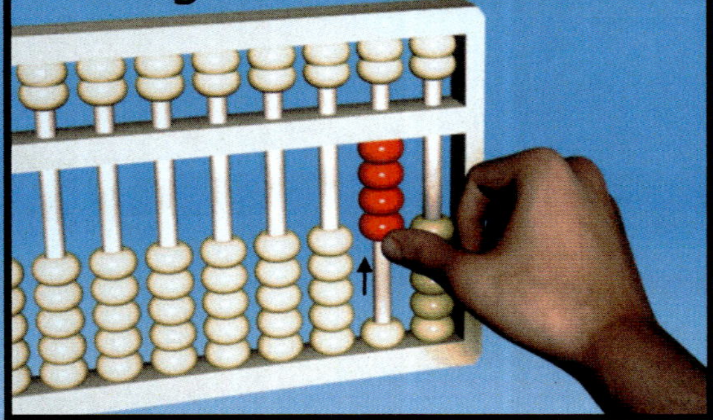

Register number 50

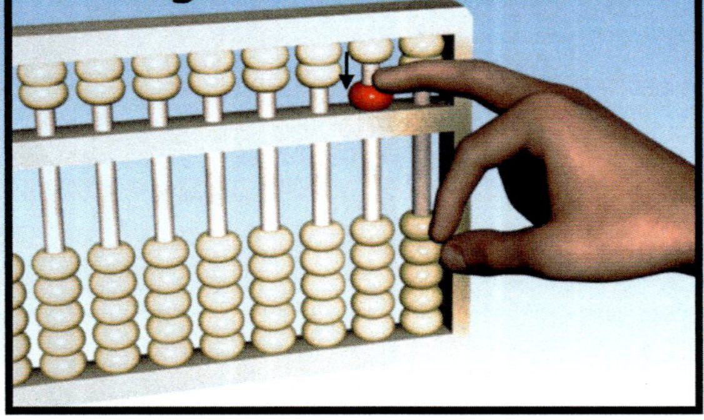

Unregister 10 from number 40

Unregister 20 from number 40

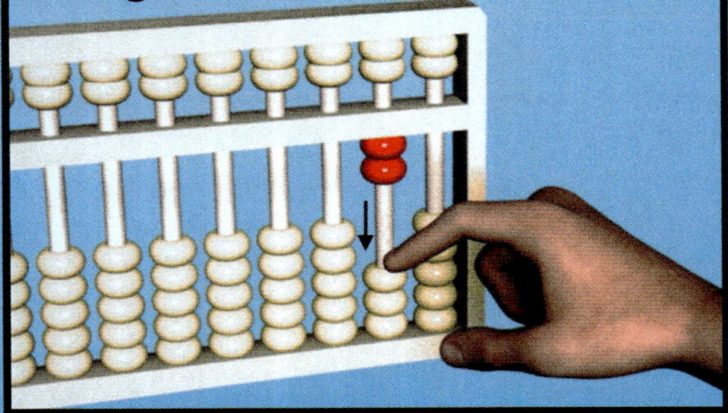

Unregister 30 from number 40

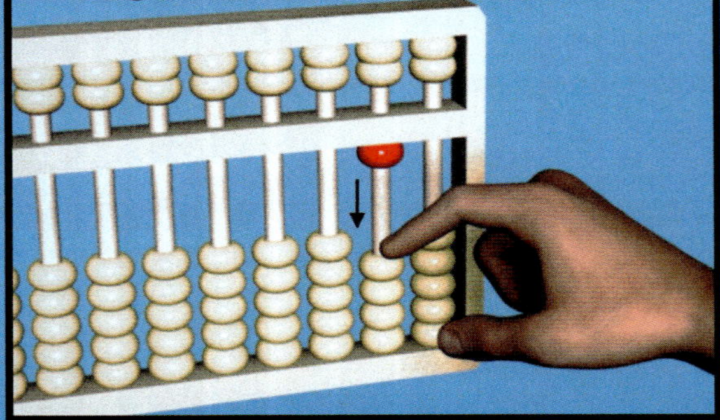

Unregister 40 from number 40

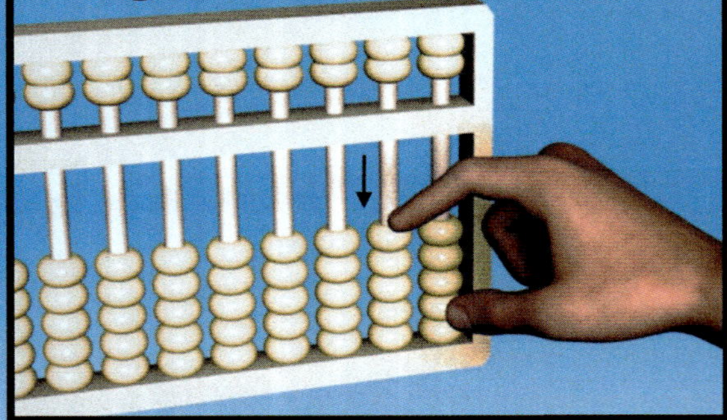

Unregister 50 from number 50

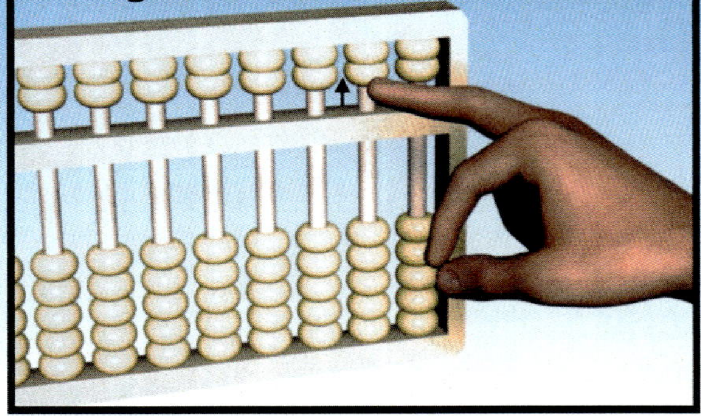

Register the number 12

First register 10

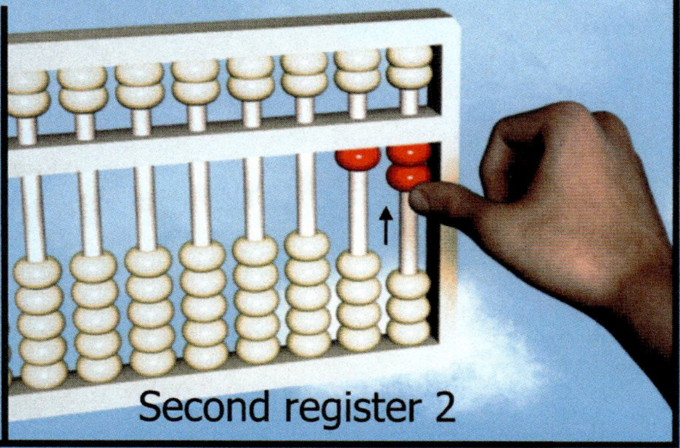
Second register 2

Unregister 12 from the number 12

First unregister 10

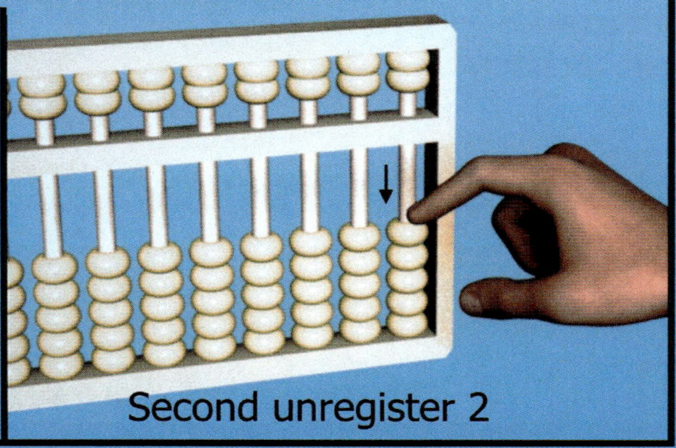
Second unregister 2

Register the number 39

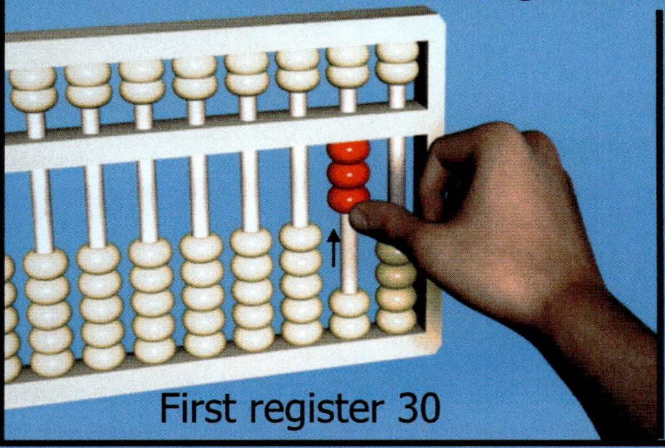
First register 30

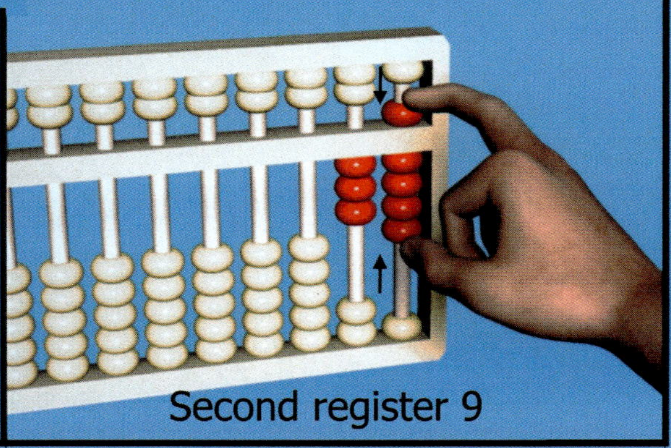
Second register 9

Unregister 12 from the number 39 = 27

First unregister 10

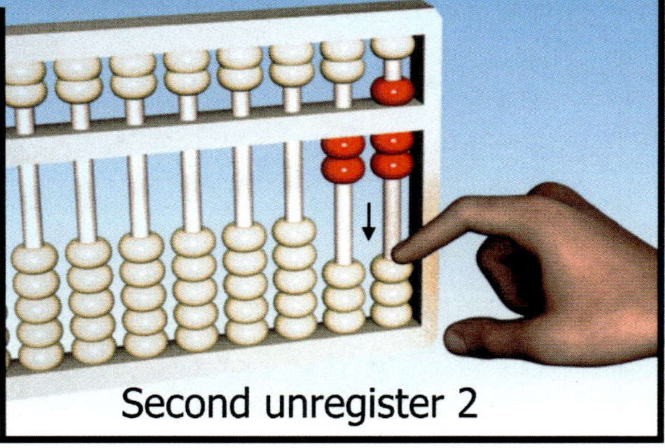
Second unregister 2

ADDITION

Addition is adding numbers to get the sum of those numbers

Addition - things to remember:

- Register your numbers from left to right, for example: for number 231 register the 2 first, 3 second and 1 last.
- Each digit must be registered in the correct column, for example with 231 the 2 is for column 3 (hundredths column), the 3 for column 2 (tens column) and the 1 for column 1 (ones column).

Example: 231 + 213

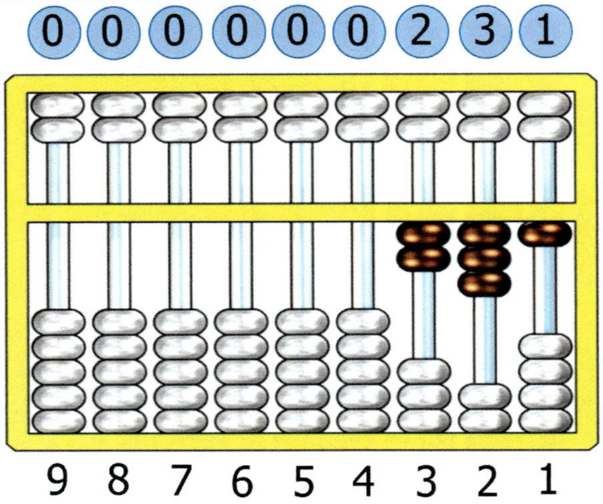

We will **register 231**

200 • Column 3, register 2 lower beads

30 • Column 2, register 3 lower beads

1 • Column 1, register 1 lower bead

The abacus reads 231

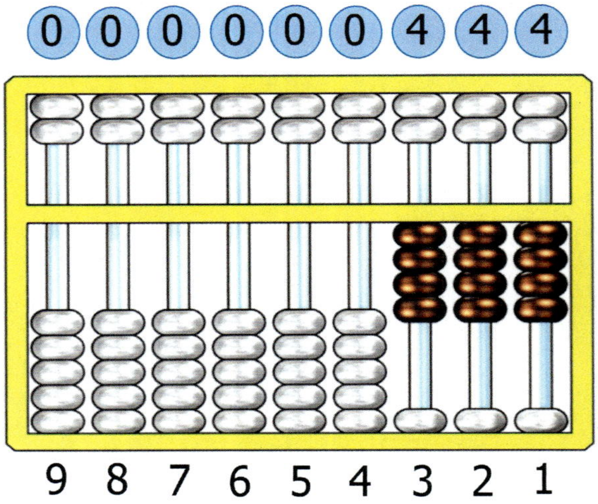

We will now **add 213** to 231

+200 • Column 3, register 2 lower beads to add 200

+10 • Column 2, register 1 lower bead to add 10

+3 • Column 1, register 3 lower beads to add 3

The abacus result is 444

These columns are useful to see the amount that you are adding.
For example:
30 means that you have just registered 30
+200 means that you have just added 200

More addition examples

Example: 4 + 5

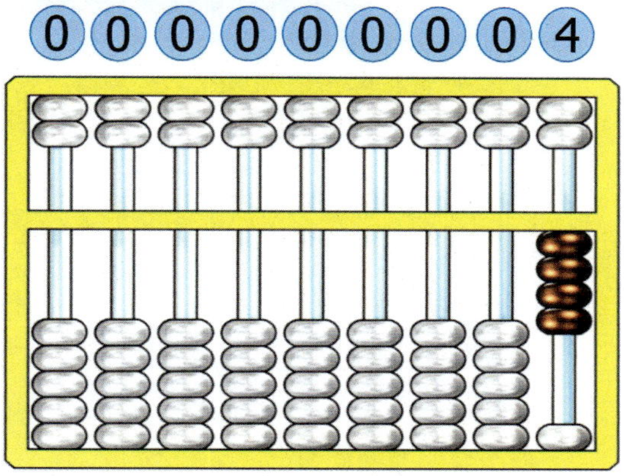

9 8 7 6 5 4 3 2 1

We will **register 4**

4 • Column 1, register 4 lower beads

The abacus reads 4

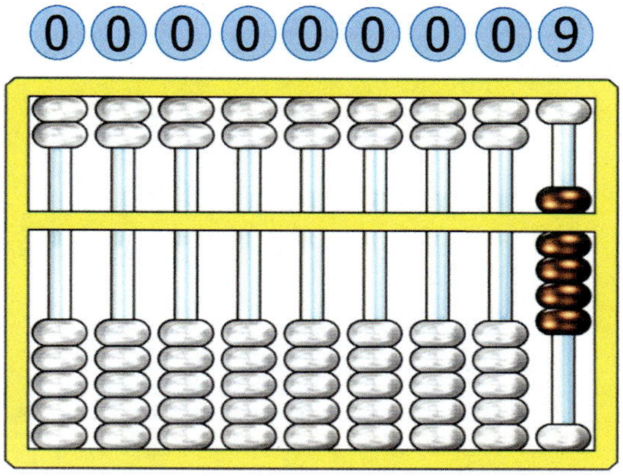

9 8 7 6 5 4 3 2 1

We will now **add 5** to 4

+5 • Column 1, register 1 upper bead to add 5

The abacus result is 9

Remember!
4 means that you have just registered 4
+5 means that you have just added 5

Example: 56 + 23

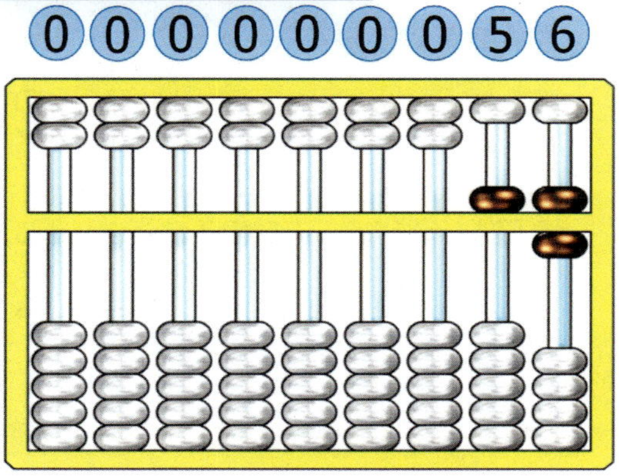

9 8 7 6 5 4 3 2 1

We will **register 56**

50 • Column 2, register 1 upper bead (this is the 5 of the 56)

6 • Column 1, register 1 upper bead and 1 lower bead (this is the 6 of the 56)

The abacus reads 56

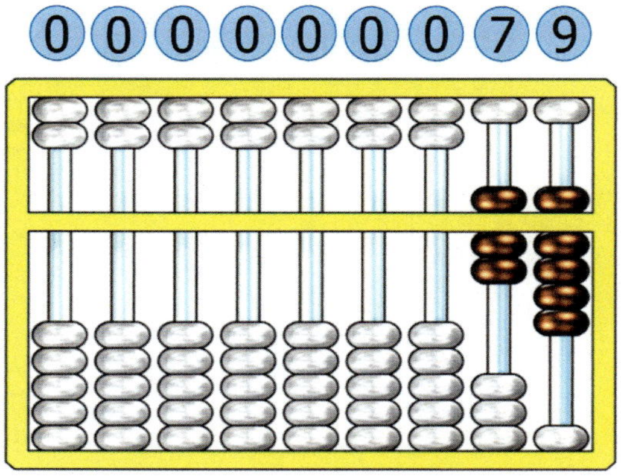

9 8 7 6 5 4 3 2 1

We will now **add 23** to 56

+20 • Column 2, register 2 lower beads to add 20

+3 • Column 1, register 3 lower beads to add 3

The abacus result is 79

More addition examples

Example: 527+ 462

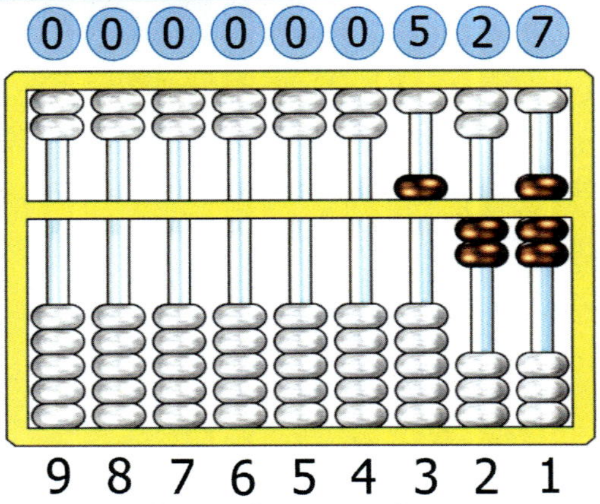

We will **register 527**

500 • Column 3, register 1 upper bead (this is the 5 of the 527)

20 • Column 2, register 2 lower beads (this is the 2 of the 527)

7 • Column 1, register 1 upper bead and 2 lower beads (this is the 7 of the 527)

The abacus reads 527

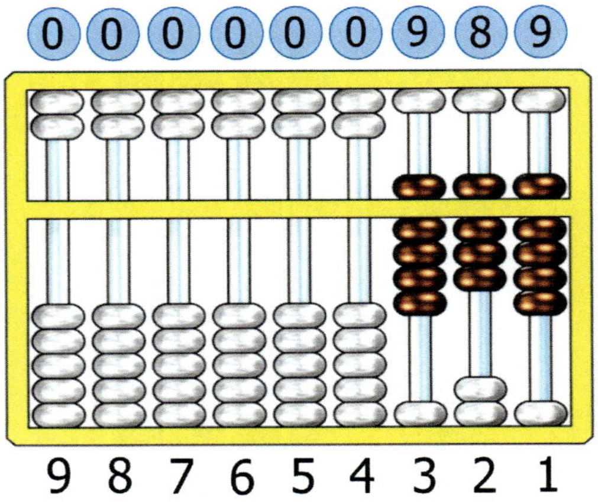

We will now **add 462** to 527

+400 • Column 3, register 4 lower beads to add 400

+60 • Column 2, register 1 upper bead and 1 lower bead to add 60

+2 • Column 1, register 2 lower beads to add 2

The abacus result is 989

Example: 2112 + 32

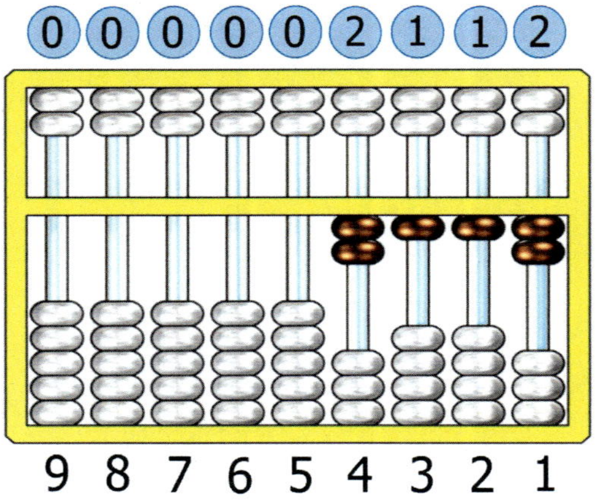

We will **register 2112**

2000 • Column 4, register 2 lower beads

100 • Column 3, register 1 lower bead

10 • Column 2, register 1 lower bead

2 • Column 1, register 2 lower beads

The abacus reads 2112

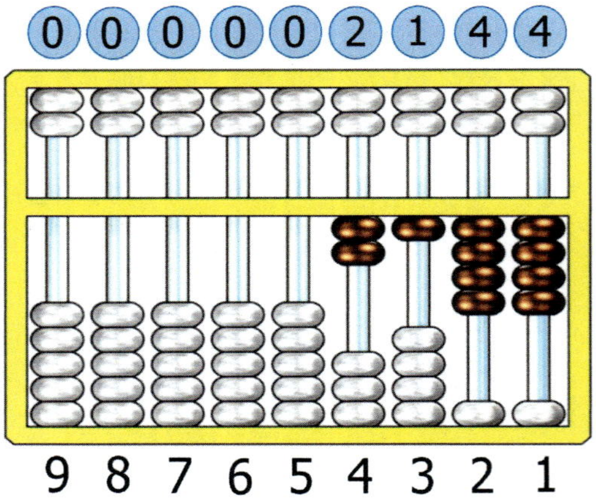

We will **add 32**

+30 • Column 2, register 3 lower beads

+2 • Column 1, register 2 lower beads

The abacus result is 2144

More addition examples

Example: 65260 + 4539

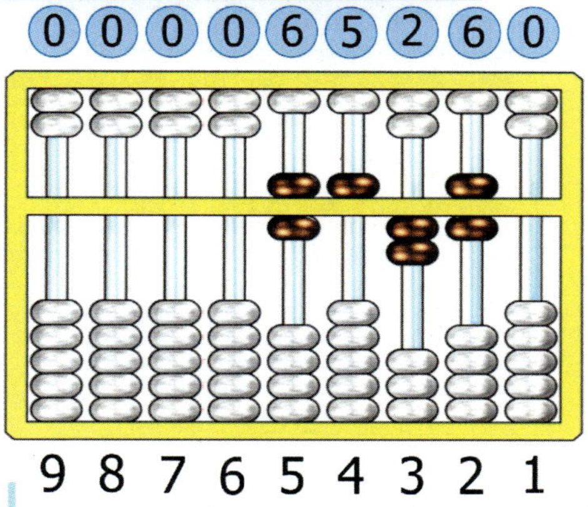

We will **register 65260**

60000 • Column 5, register 1 upper bead and 1 lower bead
5000 • Column 4, register 1 upper bead
200 • Column 3 register 2 lower beads
60 • Column 2, register 1 upper bead and 1 lower bead
• Column 1, do nothing

The abacus reads 65260

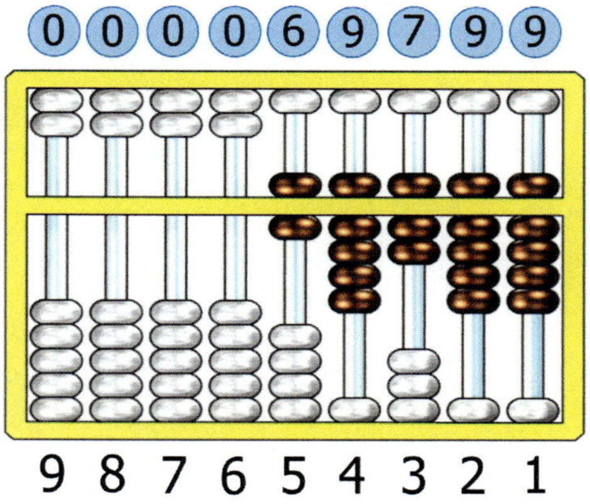

We will now **add 4539** to 65260

+4000 • Column 4, register 4 lower beads to add 4000
+500 • Column 3, register 1 upper bead to add 500
+30 • Column 2, register 3 lower beads to add 30
+9 • Column 1, register 1 upper bead and 4 lower beads to add 9

The abacus result is 69799

Example: 123456789 + 121521210

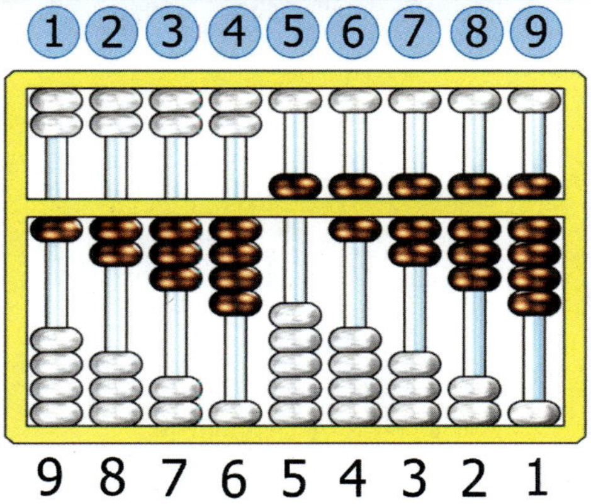

We will **register 123456789**

100000000 • Column 9, register 1 lower bead
20000000 • Column 8, register 2 lower beads
3000000 • Column 7, register 3 lower beads
400000 • Column 6, register 4 lower beads
50000 • Column 5 register 1 upper bead
6000 • Column 4, register 1 upper and 1 lower bead
700 • Column 3, register 1 upper and 2 lower beads
80 • Column 2 register 1 upper and 3 lower beads
9 • Column 1, register 1 upper and 4 lower beads

The abacus reads 123456789

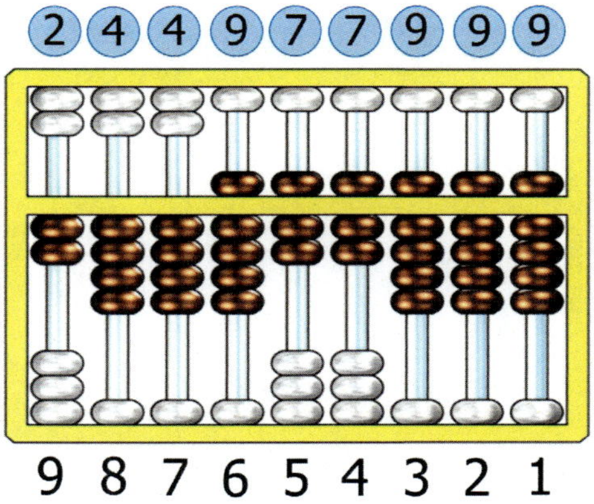

We will **add 121521210**

+100000000 • Column 9, register 1 lower bead
+20000000 • Column 8, register 2 lower beads
+1000000 • Column 7, register 1 lower bead
+500000 • Column 6, register 1 upper bead
+20000 • Column 5, register 2 lower beads
+1000 • Column 4, register 1 lower bead
+200 • Column 3, register 2 lower beads
+10 • Column 2, register 1 lower bead
• Column 1, do nothing

The abacus result is 244977999

More addition examples

Example: 523602092 + 425160905

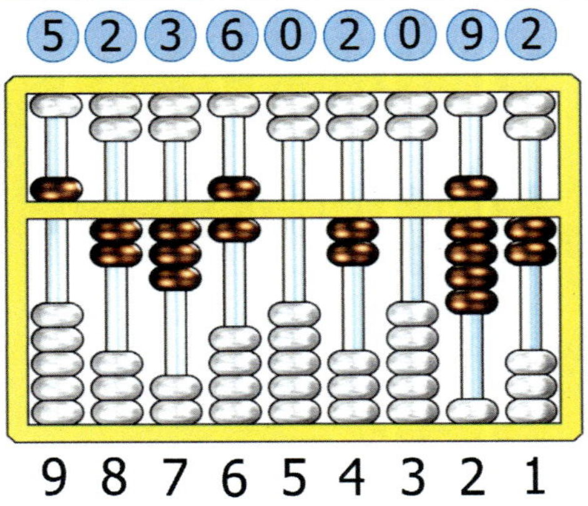

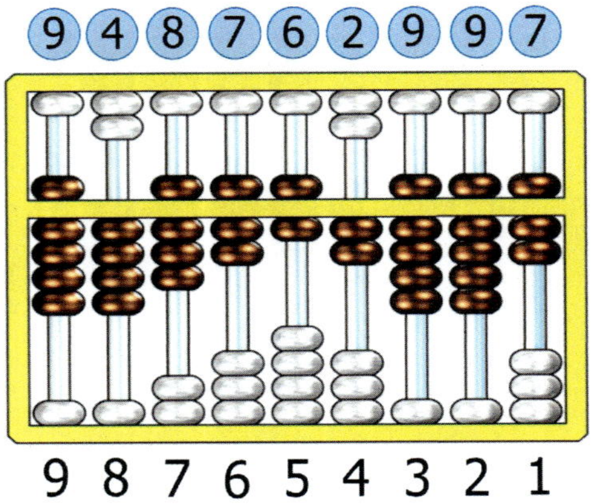

We will **register 523602092**

500000000	• Column 9, register 1 upper bead
20000000	• Column 8, register 2 lower beads
3000000	• Column 7, register 3 lower beads
600000	• Column 6, register 1 upper and 1 lower bead
	• Column 5, do nothing
2000	• Column 4, register 2 lower beads
	• Column 3, do nothing
90	• Column 2, register 1 upper and 4 lower beads
2	• Column 1, register 2 lower beads

The abacus reads 523602092

We will **add 425160905**

+400000000	• Column 9, register 4 lower beads
+20000000	• Column 8, register 2 lower beads
+5000000	• Column 7, register 1 upper bead
+100000	• Column 6, register 1 lower bead
+60000	• Column 5, register 1 upper and 1 lower bead
	• Column 4, do nothing
+900	• Column 3, register 1 upper and 4 lower beads
	• Column 2, do nothing
+5	• Column 1, register 1 upper bead

The abacus result is 948762997

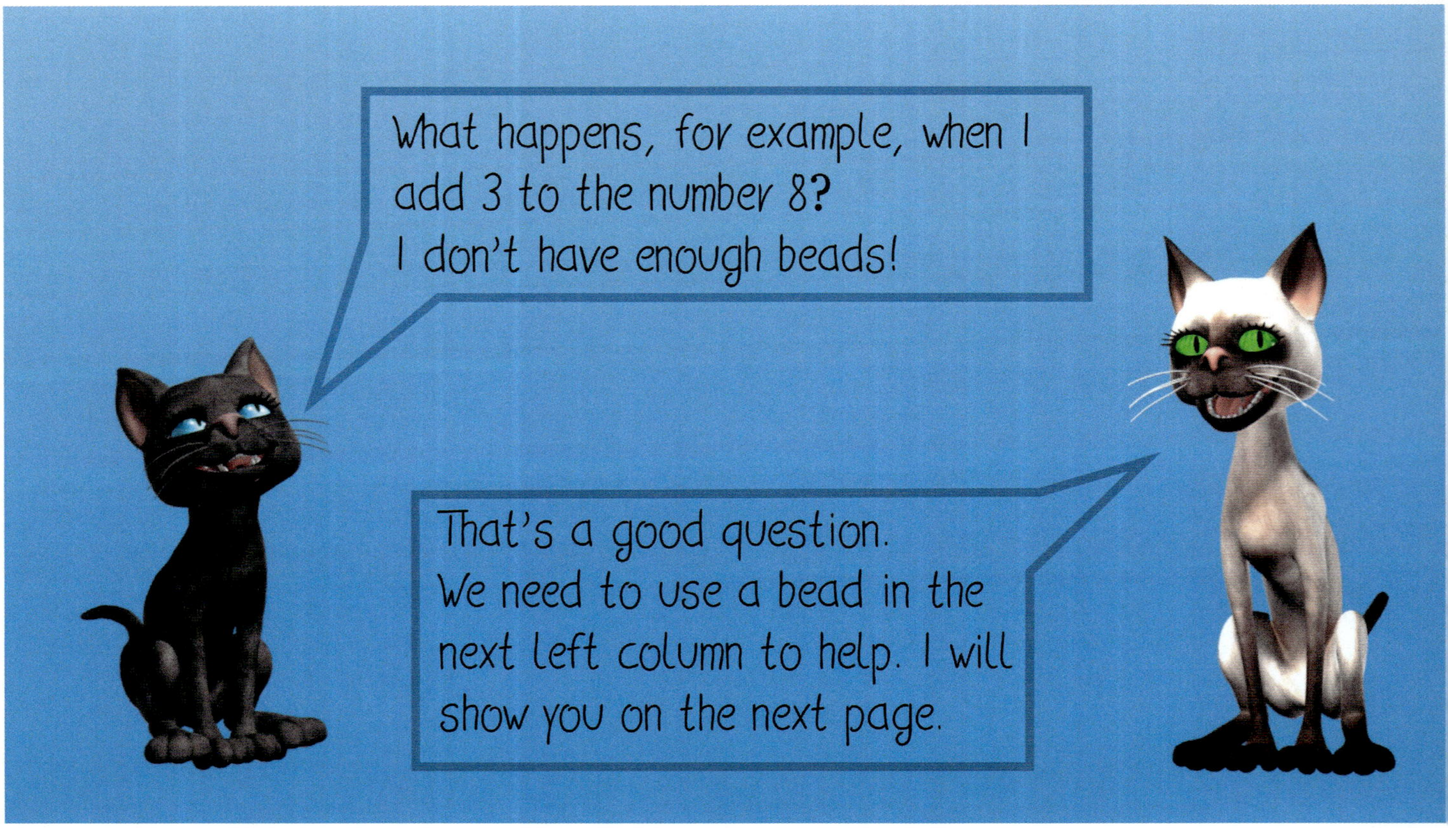

When there are not enough beads in the column for the addition

For example, when you try to add 4 to the already registered number 8, you don't have enough beads in the column to do it. You can only register a maximum of 9 in each column (4 lower beads and 1 upper bead, 4+5=9).
When this happens, we need to use the **'Not enough beads list'**.

1=10-9
2=10-8
3=10-7
4=10-6
5=10-5
6=10-4
7=10-3
8=10-2
9=10-1

How to use the 'Not enough beads list'

Let's say we need to add 3 to a column but we don't have enough beads.

Look at the list, **3=10-7**

10 is the number to **register,** in the next **LEFT** column (1 lower bead).

7 is the number to **unregister** in our column.

Example: 8 + 4

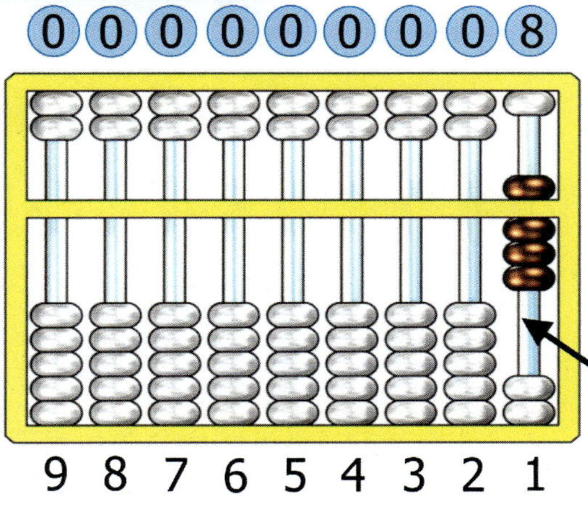

We will **register 8**

8

- Column 1, register 1 upper bead and 3 lower beads

The abacus reads 8

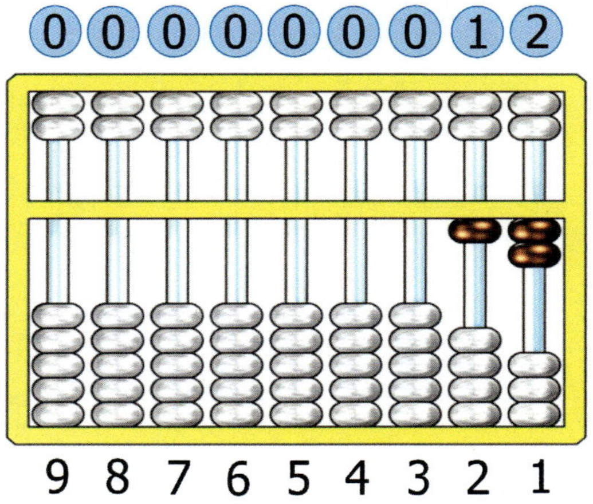

We will now **add 4**

- There are not enough beads in column 1 to add 4
- First we must think that **4=10-6** see the **'Not enough beads list'**
- -6 Column 1, **unregister** 1 upper and 1 lower bead to subtract 6 (4=10-**6**)
- +10 Column 2, register 1 lower bead to add 10 (4=**10**-6)

The abacus result is 12

More addition examples (when we don't have enough beads)

Example: 9 + 5

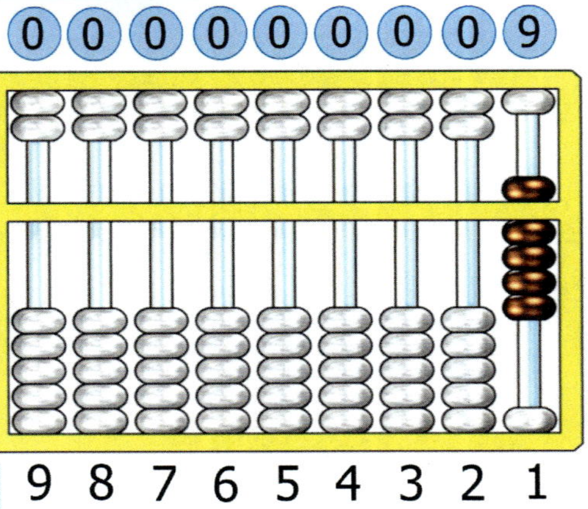

We will **register 9**

9 • Column 1, register 1 upper bead and 4 lower beads

The abacus reads 9

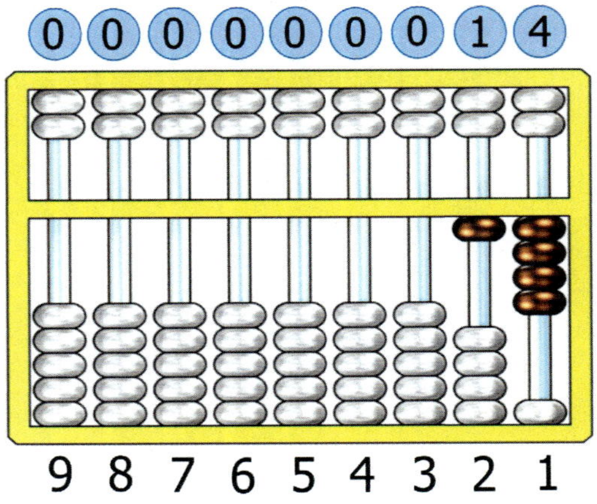

We will now **add 5**

There are not enough beads in column 1 to add 5, so think **5=10-5**, so remove 5 from column 1 then add 10 to column 2

-5 • Column 1, unregister 1 upper bead to subtract 5

+10 • Column 2, register 1 lower bead to add 10

The abacus result is 14

Example: 156 + 263

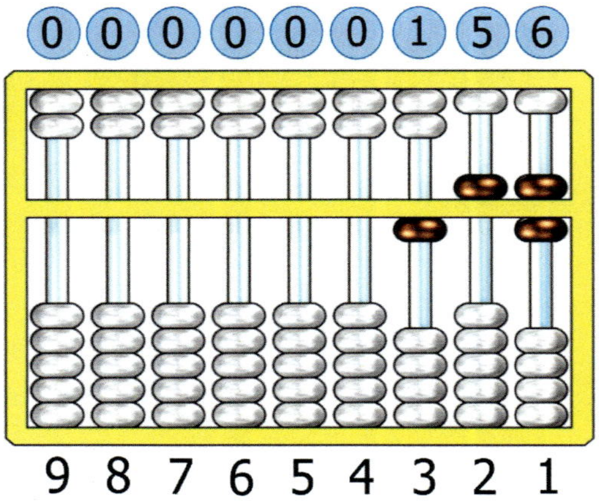

We will **register 156**

100 • Column 3, register 1 lower bead

50 • Column 2, register 1 upper bead

6 • Column 1, register 1 upper bead and 1 lower bead

The abacus reads 156

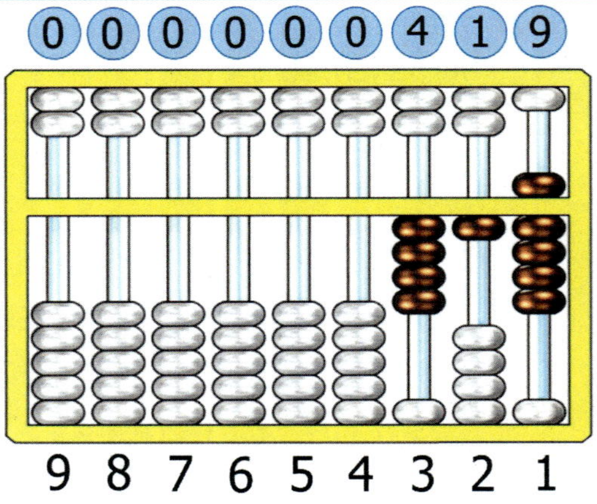

We will **add 263**

+200 • Column 3, register 2 lower beads

There are not enough beads in column 2 to register 6 more (to add 60), so think **6=10-4**

-40 • Column 2, unregister 1 upper bead (-50) and register 1 lower bead (+10)

+100 • Column 3, register 1 lower bead (Total from columns 2 & 3 is 100-50+10=60)

+3 • Column 1, register 3 lower beads

The abacus result is 419

Addition when registering and unregistering in the same column

Sometimes we need to unregister and register in the same column, as in the last example. For example if we need to add 3 beads to an already registered 4 lower beads to make 7, we need to register 1 upper and unregister 2 lower (+5-2=3).
Here are some examples.

Example: 204 + 173

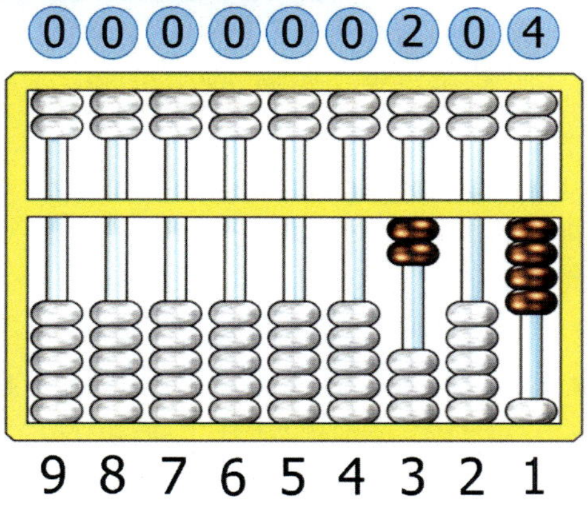

We will **register 204**

200 • Column 3, register 2 lower beads
• Column 2, do nothing
4 • Column 1, register 4 lower beads

The abacus reads 204

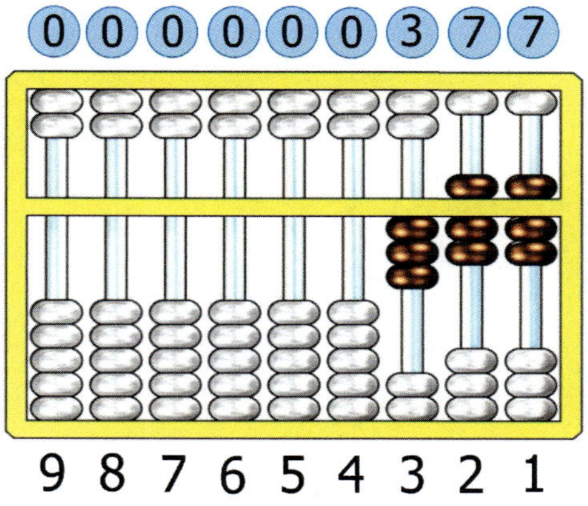

We will now **add 173**

+100 • Column 3, register 1 lower bead
+70 • Column 2, register 1 upper bead (+50) and 2 lower beads (+20)
+3 • Column 1, register 1 upper bead (+5) and unregister 2 lower beads (-2)

The abacus result is 377

Example: 6421 + 425

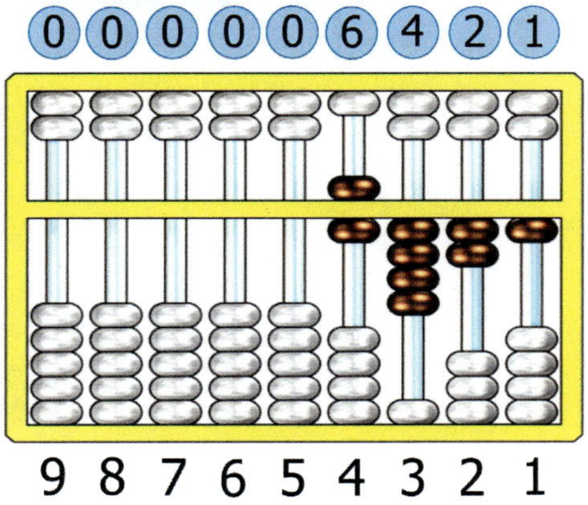

We will **register 6421**

6000 • Column 4, register 1 upper bead and 1 lower bead
400 • Column 3, register 4 lower beads
20 • Column 2, register 2 lower beads
1 • Column 1, register 1 lower bead

The abacus reads 6421

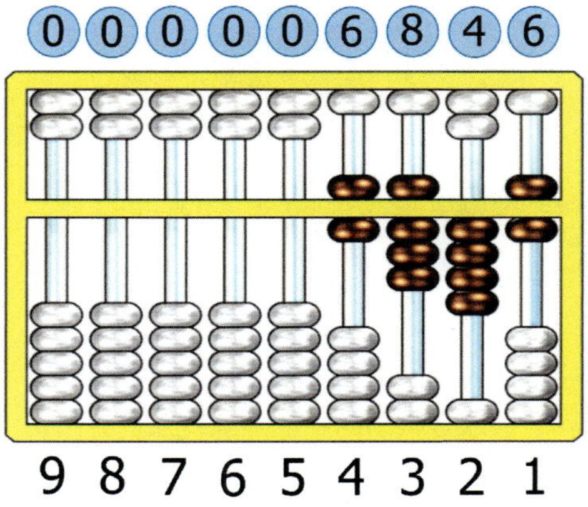

We will now **add 425**

+400 • Column 3, register 1 upper bead (+500), unregister 1 lower bead (-100)
(Total added in column 3 is 500-100=400)
+20 • Column 2, register 2 lower beads
+5 • Column 1, register 1 upper bead

The abacus result is 6846

More addition examples (when we don't have enough beads)

Example: 45 + 5

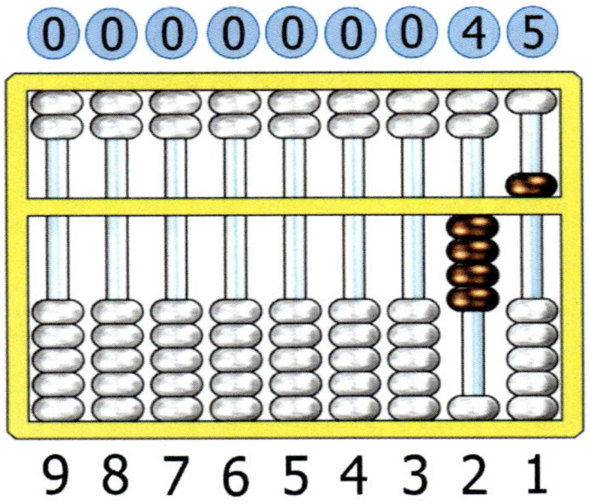

We will **register 45**

40 • Column 2, register 4 lower beads
5 • Column 1, register 1 upper bead

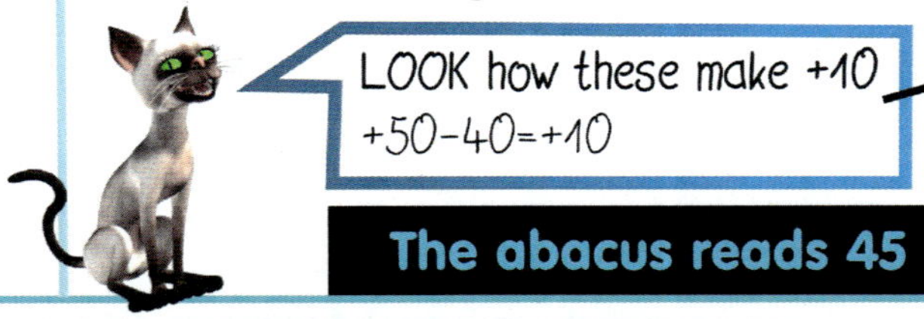

The abacus reads 45

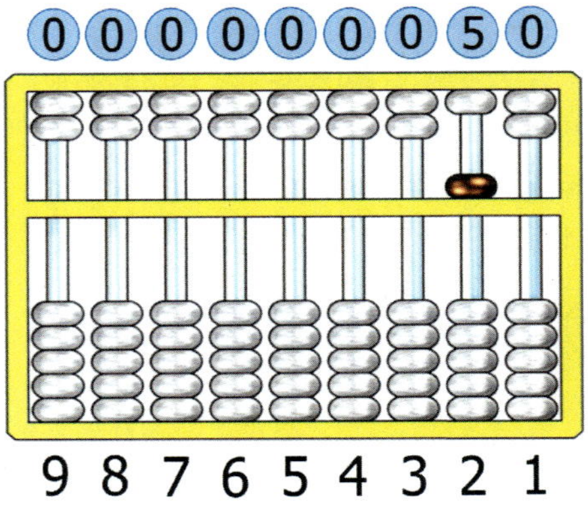

We will **add 5**

There are not enough beads in column 1 to register 5 more (to add 5), so think **5=10-5**

-5 • Column 1, unregister 1 upper bead
+50 • Column 2, register 1 upper bead
-40 • Column 2, unregister 4 lower beads

The abacus result is 50

Example: 5395607 + 2803721

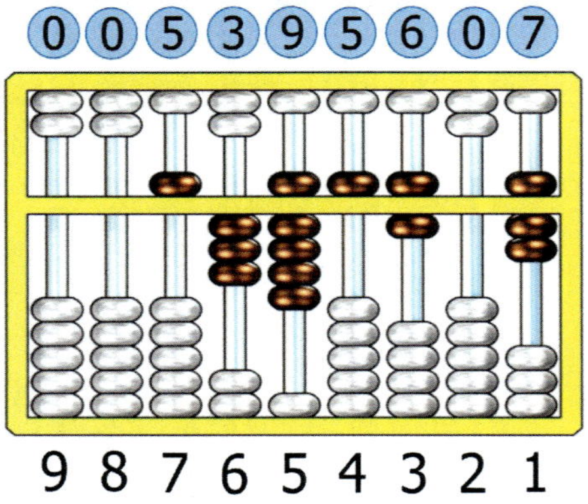

We will **register 5395607**

5000000 • Column 7, register 1 upper bead
300000 • Column 6, register 3 lower beads
90000 • Column 5, register 1 upper bead and 4 lower beads
5000 • Column 4, register 1 upper bead
600 • Column 3, register 1 upper bead and 1 lower bead
• Column 2, do nothing
7 • Column 1, register 1 upper bead and 2 lower beads

The abacus reads 5395607

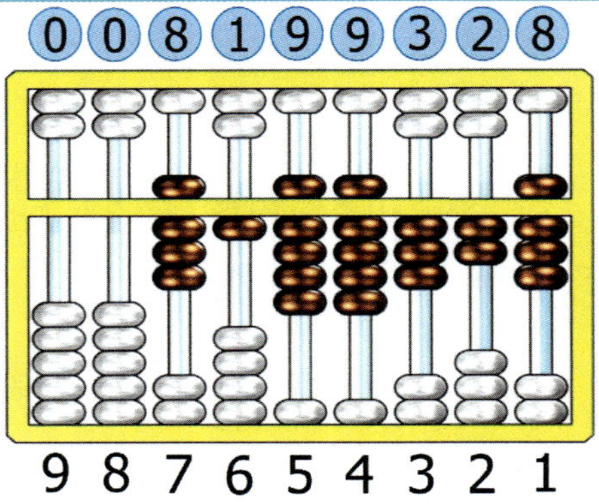

We will **add 2803721**

+2000000 • Column 7, register 2 lower beads

There are not enough beads in column 6 to register 8 more, so think **8=10-2**

-200000 • Column 6, unregister 2 lower beads
+1000000 • Column 7, register 1 lower bead

• Column 5, do nothing
+3000 • Column 4, register 3 lower beads

There are not enough beads in column 3 to register 7 more, so think **7=10-3**

-500 • Column 3, unregister 1 upper bead
+200 • Column 3, register 2 lower beads
+1000 • Column 4, register 1 lower bead

+20 • Column 2, register 2 lower beads
+1 • Column 1, register 1 lower bead

The abacus result is 8199328

More addition examples (when we don't have enough beads)

Example: 99999 + 11111

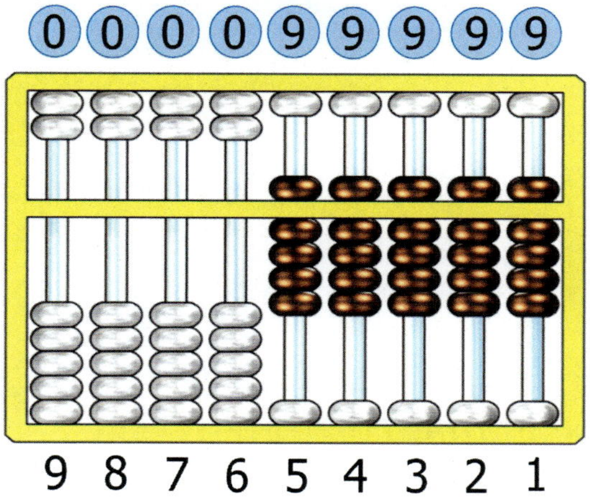

We will **register 99999**

- 90000 • Column 5, register 1 upper bead and 4 lower beads
- 9000 • Column 4, register 1 upper bead and 4 lower beads
- 900 • Column 3, register 1 upper bead and 4 lower beads
- 90 • Column 2, register 1 upper bead and 4 lower beads
- 9 • Column 1, register 1 upper bead and 4 lower beads

Can you see a pattern?

LOOK how these make +100
+1000-900=+100

LOOK how these make +10
+100-90=+10

LOOK how these make +1
+10-9=+1

The abacus reads 99999

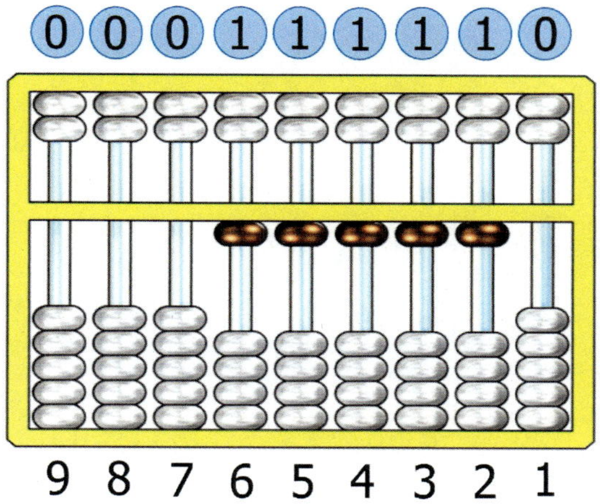

We will **add 11111**

There are not enough beads in column 5 to register 1 more, so think **1=10-9**
- -90000 • Column 5, unregister 1 upper bead and 4 lower beads
- +100000 • Column 6, register 1 lower bead

There are not enough beads in column 4 to register 1 more, so think **1=10-9**
- -9000 • Column 4, unregister 1 upper bead and 4 lower beads
- +10000 • Column 5, register 1 lower bead

There are not enough beads in column 3 to register 1 more, so think **1=10-9**
- -900 • Column 3, unregister 1 upper bead and 4 lower beads
- +1000 • Column 4, register 1 lower bead

There are not enough beads in column 2 to register 1 more, so think **1=10-9**
- -90 • Column 2, unregister 1 upper bead and 4 lower beads
- +100 • Column 3, register 1 lower bead

There are not enough beads in column 1 to register 1 more, so think **1=10-9**
- -9 • Column 1, unregister 1 upper bead and 4 lower beads
- +10 • Column 2, register 1 lower bead

The abacus result is 111110

Skipped columns when adding

Sometimes we have to SKIP a column. I'll explain why below.

We've learnt that when a column doesn't have enough beads left on it to make the addition, we move to the next LEFT column to help. Sometimes the next left column also doesn't have enough beads on it, so we **SKIP** this column and move again to the next left column until you reach a column that has enough beads to use. See below how it works.

Important!

- We will **SKIP** a column when there are not enough beads to use in that column.
- We will see this symbol when we need to skip a column (move on to the next left column).

←

- We will **UNREGISTER** all beads in any skipped columns.

Example: 95 + 8

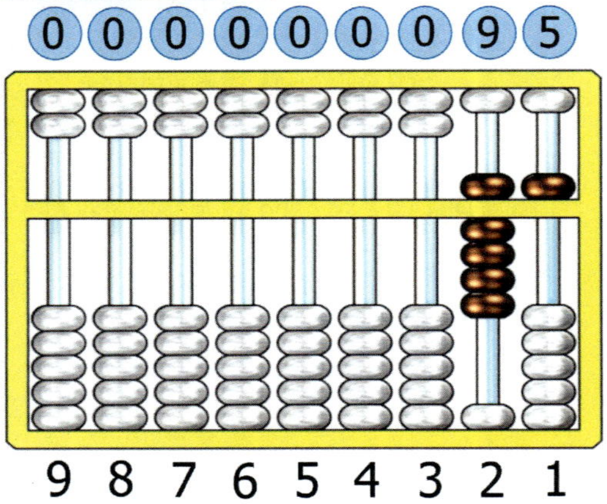

We will **register 95**

90 • Column 2, register 1 upper bead and 4 lower beads

5 • Column 1, register 1 upper bead

This big arrow means SKIP

LOOK how these make +10
+100−90=+10

The abacus reads 95

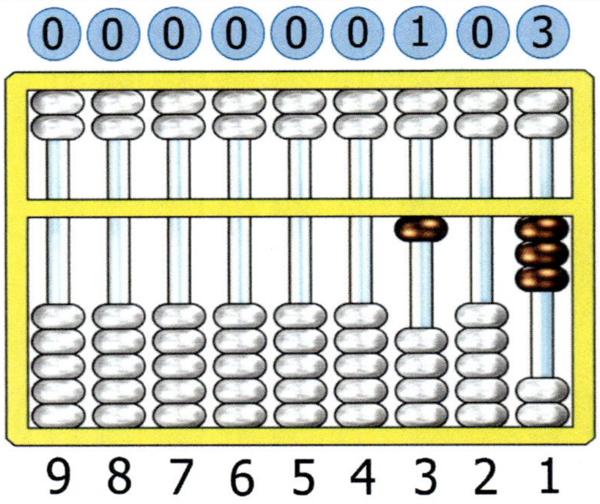

We will now **add 8**

There are not enough beads in column 1 to register 8 more, so think **8=10-2**

-2 • Column 1, unregister 1 upper bead and register 3 lower beads

We would normally move to the next LEFT column and add 1 lower bead (to add 10) but we don't have 1 bead left to use

← • Column 2, **SKIP** this column and unregister all beads

-90

+100 • Column 3, register 1 lower bead

The abacus result is 103

More skipped column examples

Example: 995 + 9

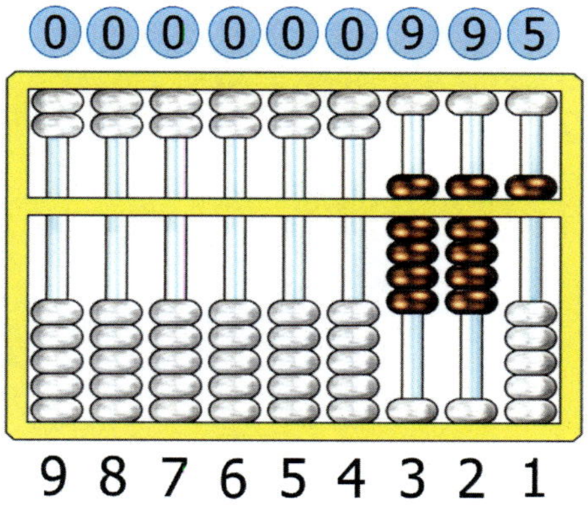

We will **register 995**

900 • Column 3, register 1 upper bead and 4 lower beads

90 • Column 2, register 1 upper bead and 4 lower beads

5 • Column 1, register 1 upper bead

The abacus reads 995

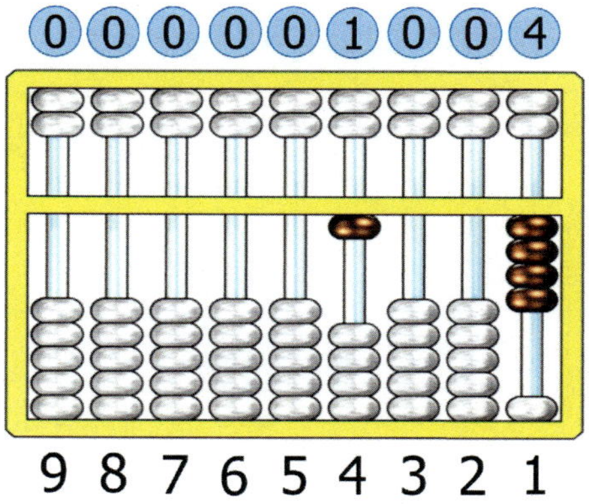

We will now **add 9**

There are not enough beads in column 1 to register 9 more, so think **9=10-1**

-1 • Column 1, unregister 1 upper bead and register 4 lower beads

← • Column 2, **SKIP** this column and

-90 unregister all beads

← • Column 3, **SKIP** this column and

-900 unregister all beads

+1000 • Column 4, register 1 lower bead

The abacus result is 1004

Example: 9999 + 1

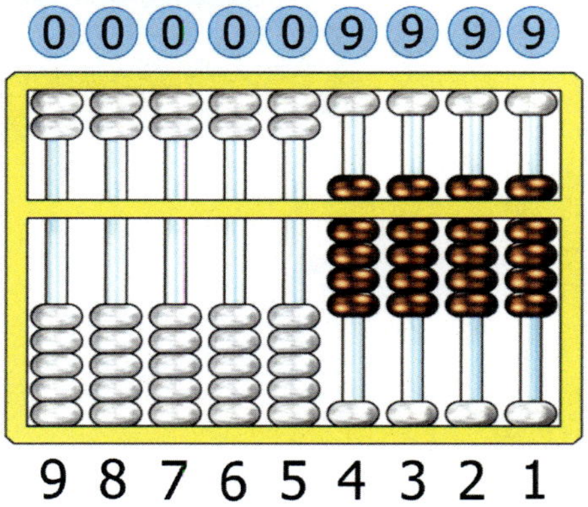

We will **register 9999**

9000 • Column 4, register 1 upper bead and 4 lower beads

900 • Column 3, register 1 upper bead and 4 lower beads

90 • Column 2, register 1 upper bead and 4 lower beads

9 • Column 1, register 1 upper bead and 4 lower beads

The abacus reads 9999

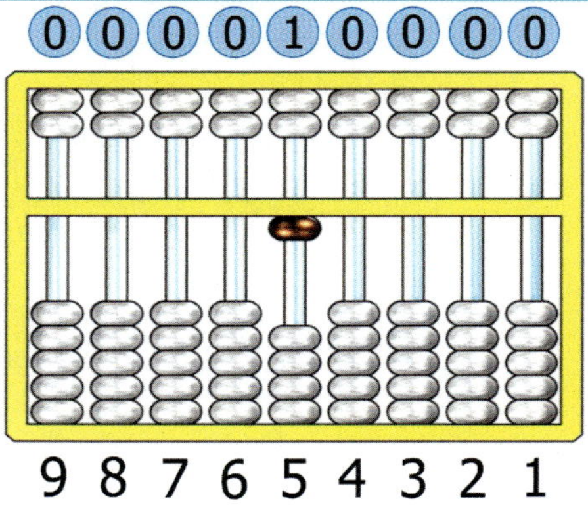

We will now **add 1**

There are not enough beads in column 1 to register 1 more, so think **1=10-9**

-9 • Column 1, unregister all beads

← • Column 2, **SKIP** this column and

-90 unregister all beads, move to column 3

← • Column 3, **SKIP** this column and

-900 unregister all beads, move to column 4

← • Column 4, **SKIP** this column and

-9000 unregister all beads, move to column 5

+10000 • Column 5, register 1 lower bead

The abacus result is 10000

Addition of more than two numbers

Sometimes we have to add 3 or more numbers, here's how.

When we add many numbers on the abacus, just find the sum of the first two, then add the next number to that sum.

Keep adding one number to the sum of the previous numbers until all the numbers have been added.

Example: 123 + 254 + 522

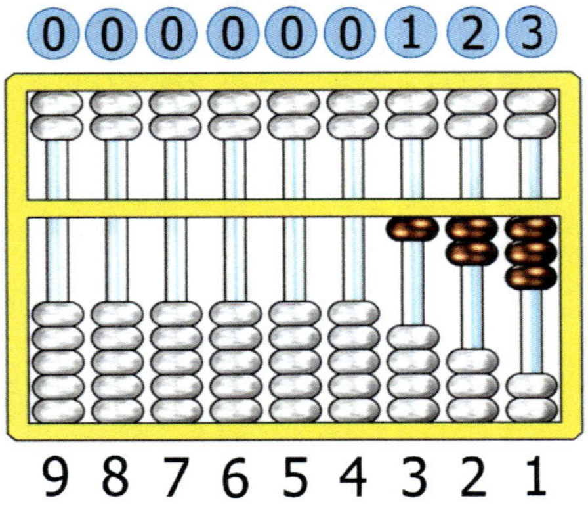

We will **register 123**

100 • Column 3, register 1 lower bead
20 • Column 2, register 2 lower beads
3 • Column 1, register 3 lower beads

The abacus reads 123

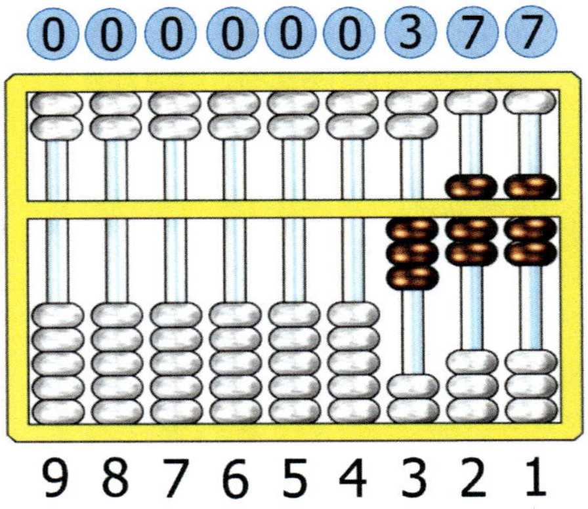

We will now **add 254 to 123**

+200 • Column 3, register 2 lower beads
+50 • Column 2, register 1 upper bead
+4 • Column 1, register 1 upper bead and unregister 1 lower bead

The abacus sum is 377

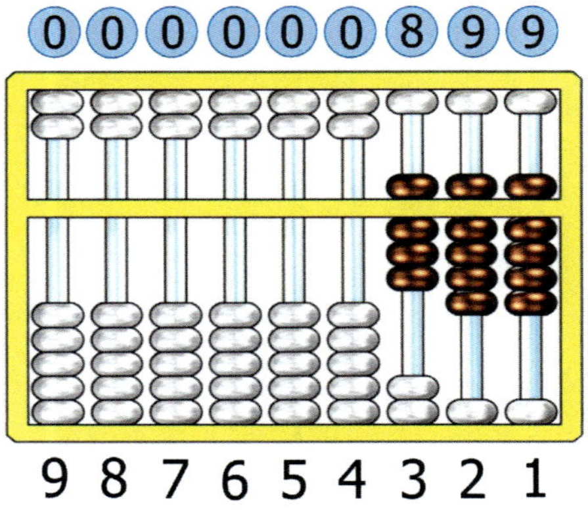

We will now **add 522 to the sum 377**

+500 • Column 3, register 1 upper bead
+20 • Column 2, register 2 lower beads
+2 • Column 1, register 2 lower beads

The abacus result is 899

Addition of more than two numbers

Example: 525631 + 253160 + 1210

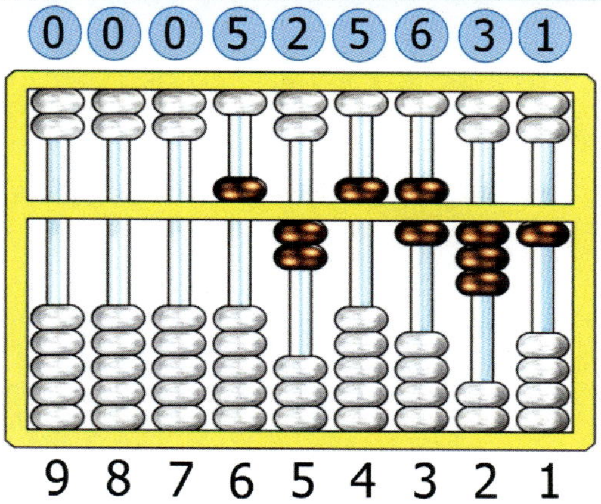

We will **register 525631**

500000	• Column 6, register 1 upper bead
20000	• Column 5, register 2 lower beads
5000	• Column 4, register 1 upper bead
600	• Column 3, register 1 upper bead and 1 lower bead
30	• Column 2, register 3 lower beads
1	• Column 1, register 1 lower bead

The abacus reads 525631

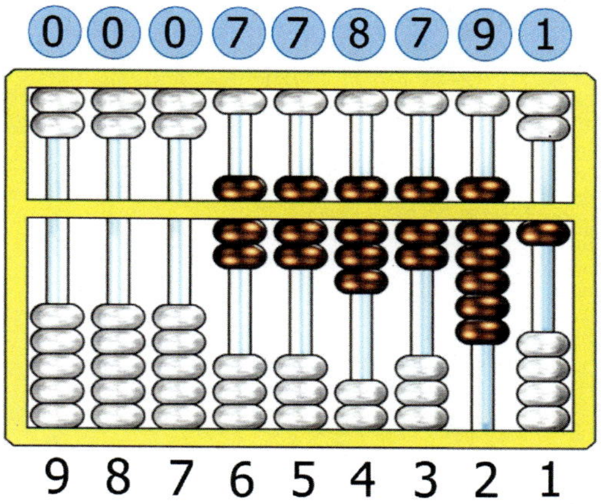

We will now **add 253160 to 525631**

+200000	• Column 6, register 2 lower beads
+50000	• Column 5, register 1 upper bead
+3000	• Column 4, register 3 lower beads
+100	• Column 3, register 1 lower bead
+60	• Column 2, register 1 upper bead and 1 lower bead
	• Column 1, do nothing

The abacus sum is 778791

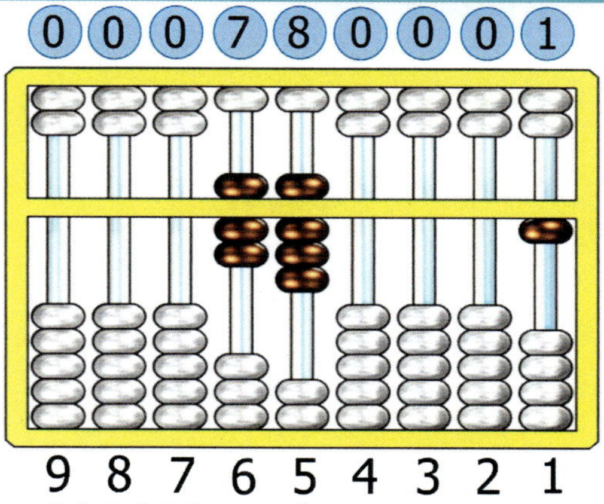

After we registered these, columns 4 & 3 had no beads left to use.

We will now **add 1210 to the sum 778791**

1000	• Column 4, register 1 lower bead
200	• Column 3, register 2 lower beads
	There are not enough beads in column 2 to register 1 more, so think **1=10-9**
-90	• Column 2,unregister all beads
← -900	• Column 3, **SKIP** this column and unregister all beads, move to column 4
← -9000	• Column 4, **SKIP** this column and unregister all beads, move to column 5
+10000	• Column 5, register 1 lower bead
	• Column 1, do nothing

The abacus result is 780001

TEST 2 - Addition

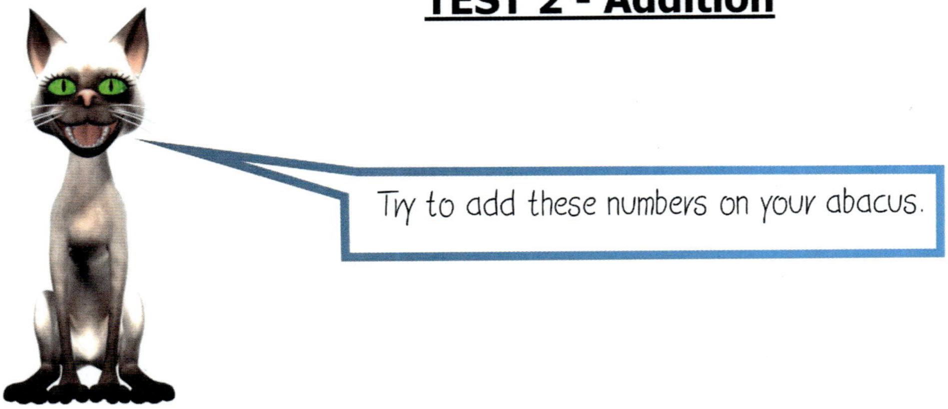

Answers are on page 44 to 43.

1. 58 + 45
2. 564 + 135
3. 885 + 9
4. 5253 + 3231
5. 55227 + 11111
6. 732689 + 555201
7. 4321543 + 5365
8. 1535 + 252 + 22
9. 135254 + 2560 + 125 + 52151
10. 12345678 + 10000009 + 91255450

Answers to test 2 (on page 40)

Here are the addition answers.
If you got any wrong, just have another go.

1

58 + 45 = **103**

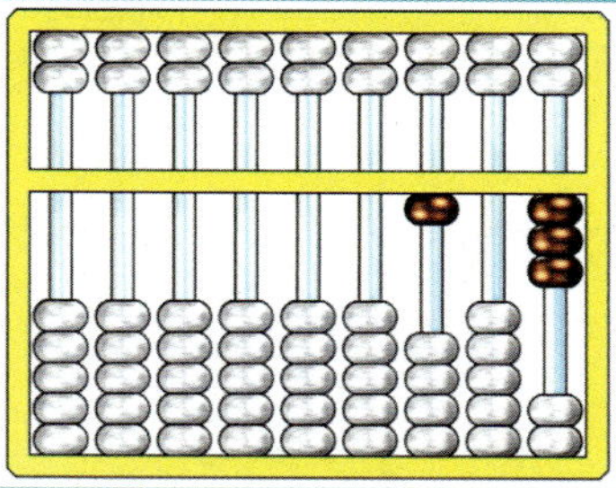

2

564 + 135 = **699**

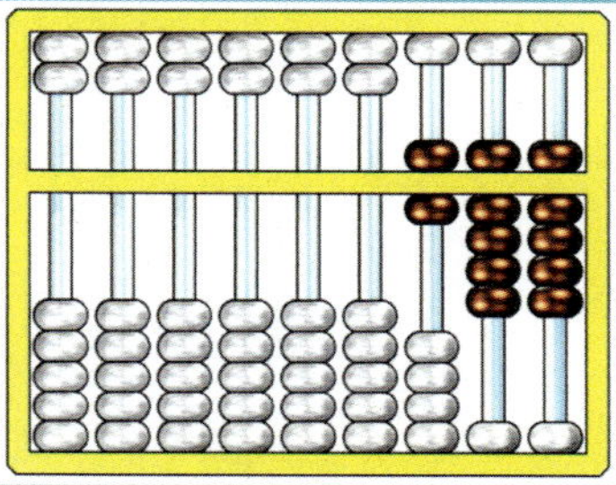

3

885 + 9 = **894**

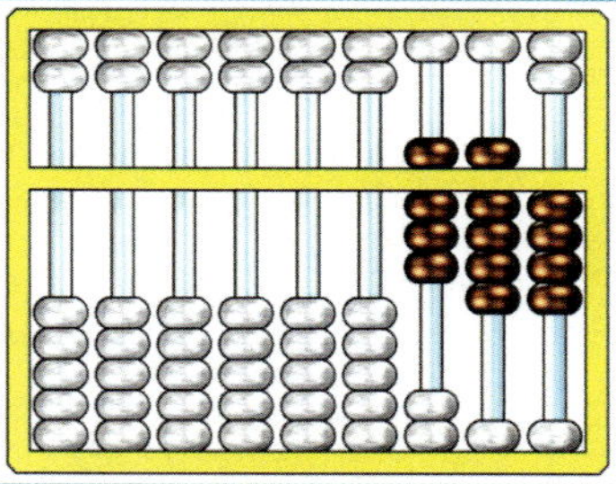

4

5253 + 3231 = **8484**

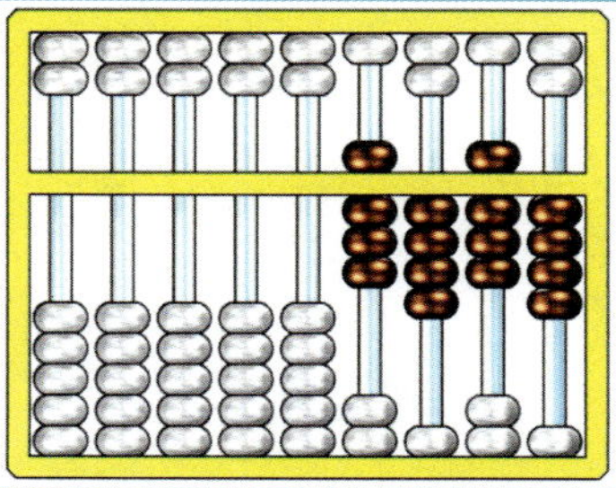

5

55227 + 11111 = **66338**

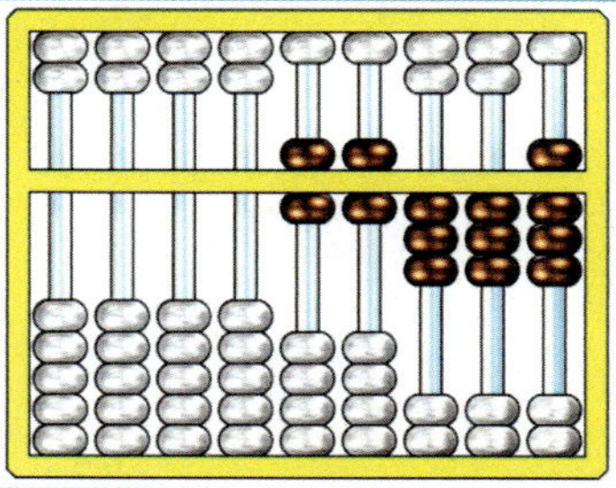

Answers to test 2 (on page 40)

6 732689 + 555201 = **1287890**

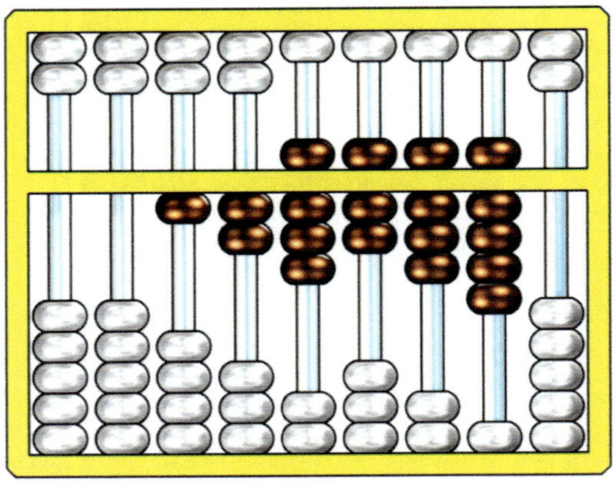

7 4321543 + 5365 = **4326908**

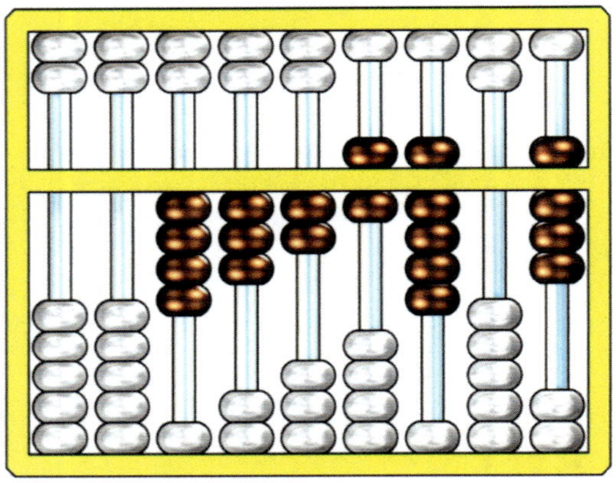

8 1535 + 252 + 22 = **1809**

1535 + 252 = **1787**

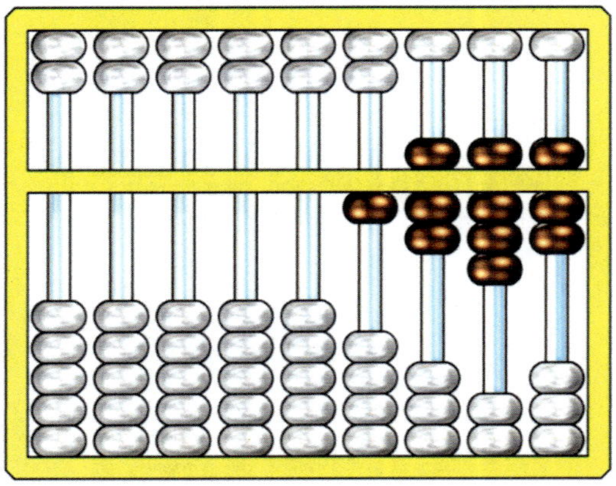

1787 + 22 = **1809**

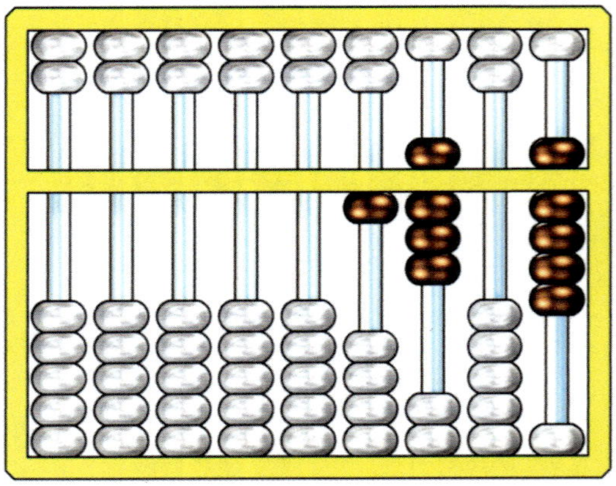

Answers to test 2 (on page 40)

9 135254 + 2560 + 125 + 52151 = **190090**

135254 + 2560 = **137814**

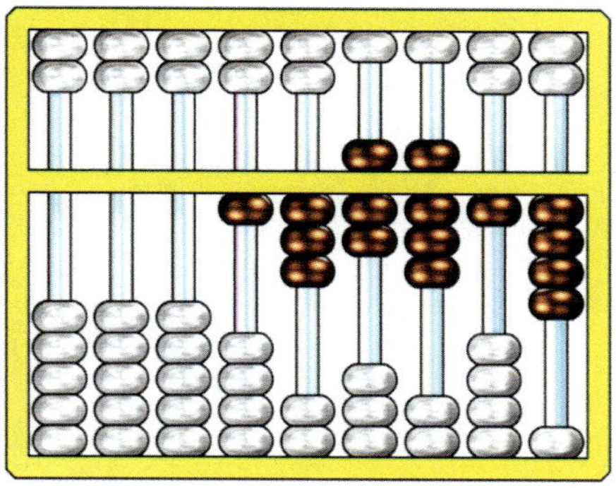

137814 + 125 = **137939**

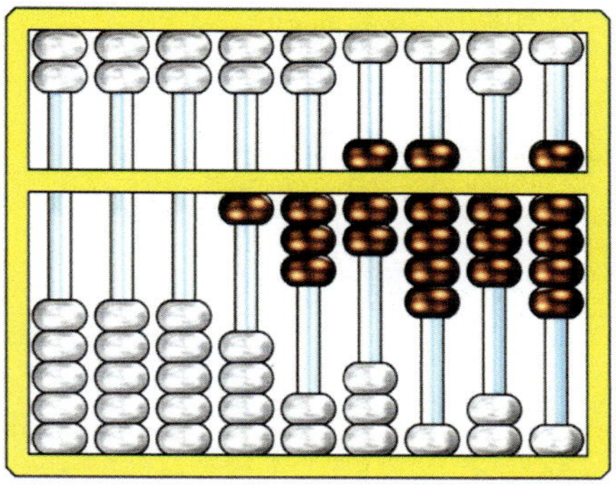

137939 + 52151 = **190090**

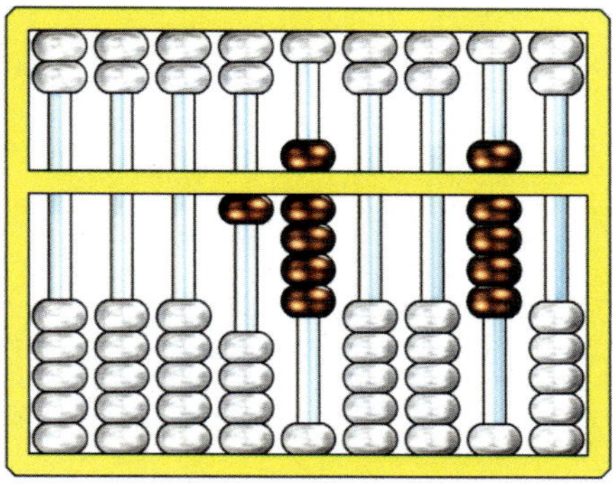

10 12345678 + 10000009 + 91255450 = **113601137**

12345678 + 10000009 = **22345687**

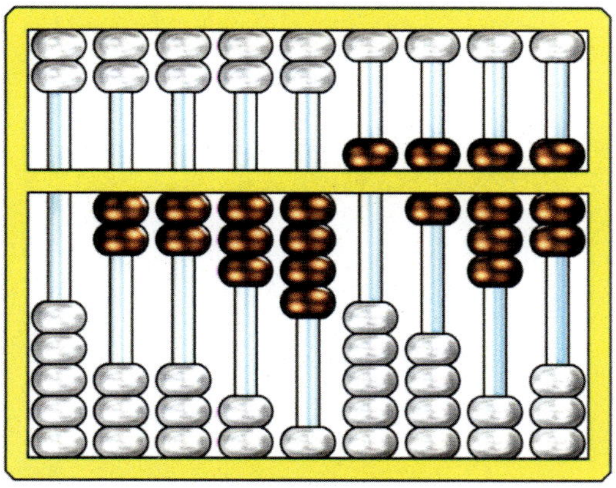

22345687 + 91255450 = **113601137**

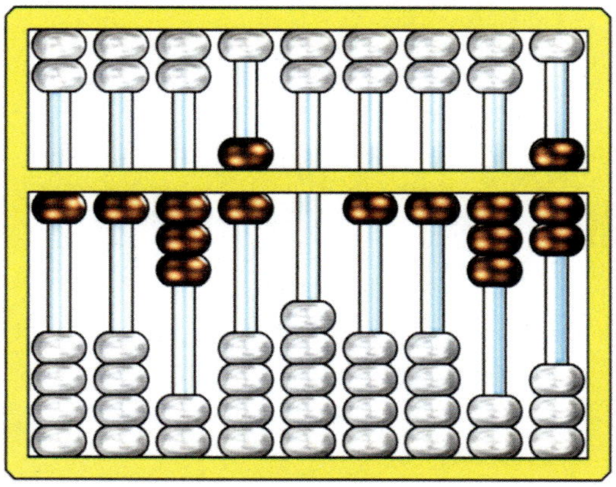

Addition of decimal numbers

A decimal number is a number that contains a decimal point, like 0.5 and it is a part of a whole number.

When we put a decimal number on the abacus, we need to draw or stick a dot on the beam to be our decimal point. In our case, any number to the right of the dot (on columns 1, 2 and 3) will be our decimals because we have a dot before column 4. Look at these examples:

Example: 1.87

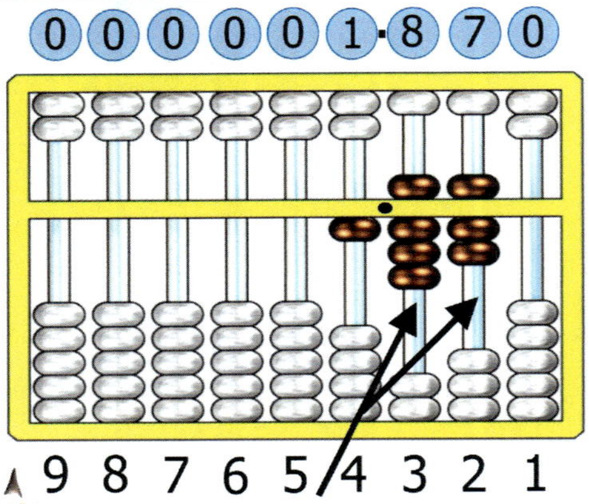

Look how the 8 of the 1.87 is in the 3rd column & the 7 of the 1.87 in in the 2nd column!

Example: 0.14

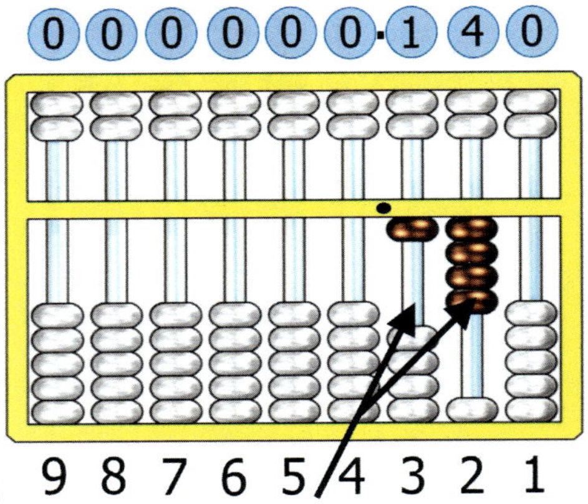

Look how the 1 of the 0.14 is in the 3rd column and the 4 of the 0.14 is in the 2nd column!

Example: 3.213 + 1.22

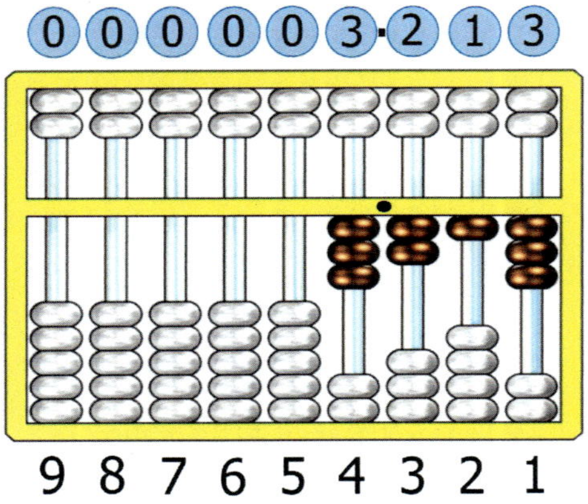

We will **register 3.213**

3	• Column 4, register 3 lower beads
0.2	• Column 3, register 2 lower beads
0.01	• Column 2, register 1 lower bead
0.003	• Column 1, register 3 lower beads

The abacus reads 3.213

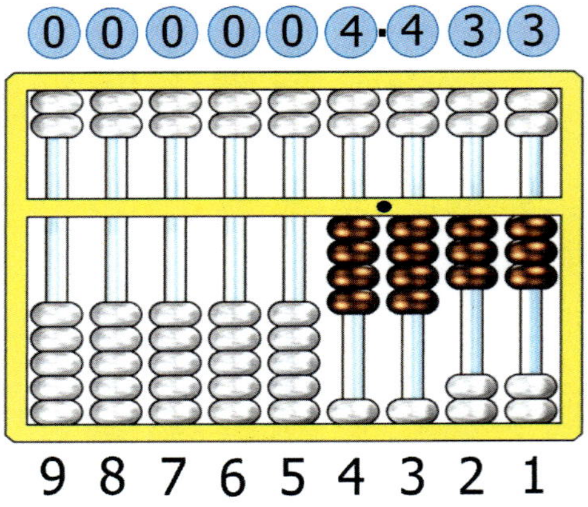

We will **add 1.22**

+1	• Column 4, register 1 lower bead
+0.2	• Column 3, register 2 lower beads
+0.02	• Column 2, register 2 lower beads
	• Column 1, do nothing

The abacus result is 4.433

More addition of decimal number examples

Example: 6.421 + 0.659

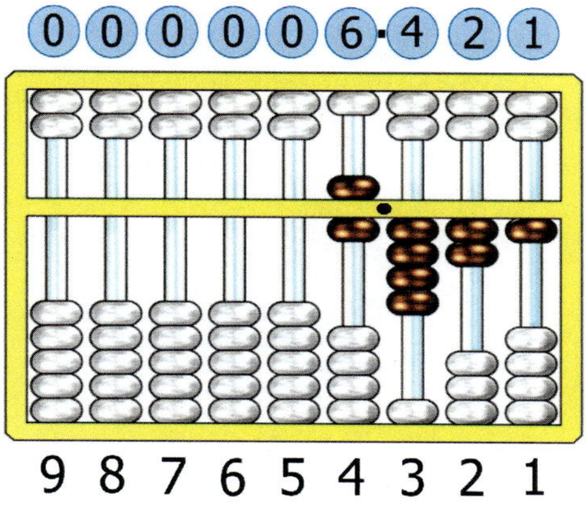

We will **register 6.421**

6	• Column 4, register 1 upper bead and 1 lower bead
0.4	• Column 3, register 4 lower beads
0.02	• Column 2, register 2 lower beads
0.001	• Column 1, register 1 lower bead

The abacus reads 6.421

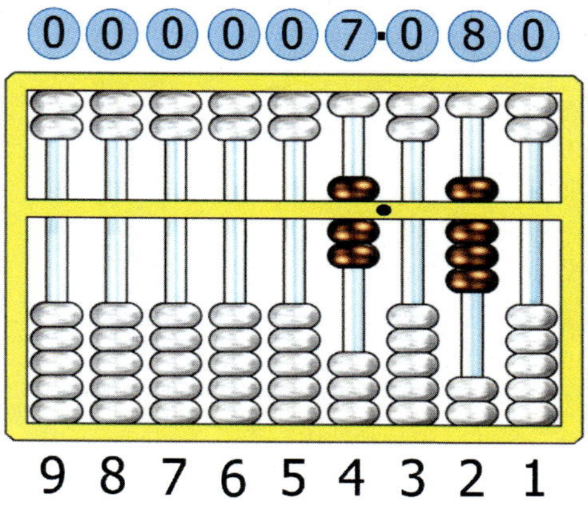

We will now **add 0.659**

	• Column 4, do nothing
	There are not enough beads in column 3 to register 6 more, so think **6=10-4**
-0.4	• Column 3, unregister 4 lower beads
+1	• Column 4, register 1 lower bead
+0.05	• Column 2, register 1 upper bead
	There are not enough beads in column 1 to register 9 more, so think **9=10-1**
-0.001	• Column 1, unregister 1 lower bead
+0.01	• Column 2, register 1 lower bead

The abacus result is 7.08

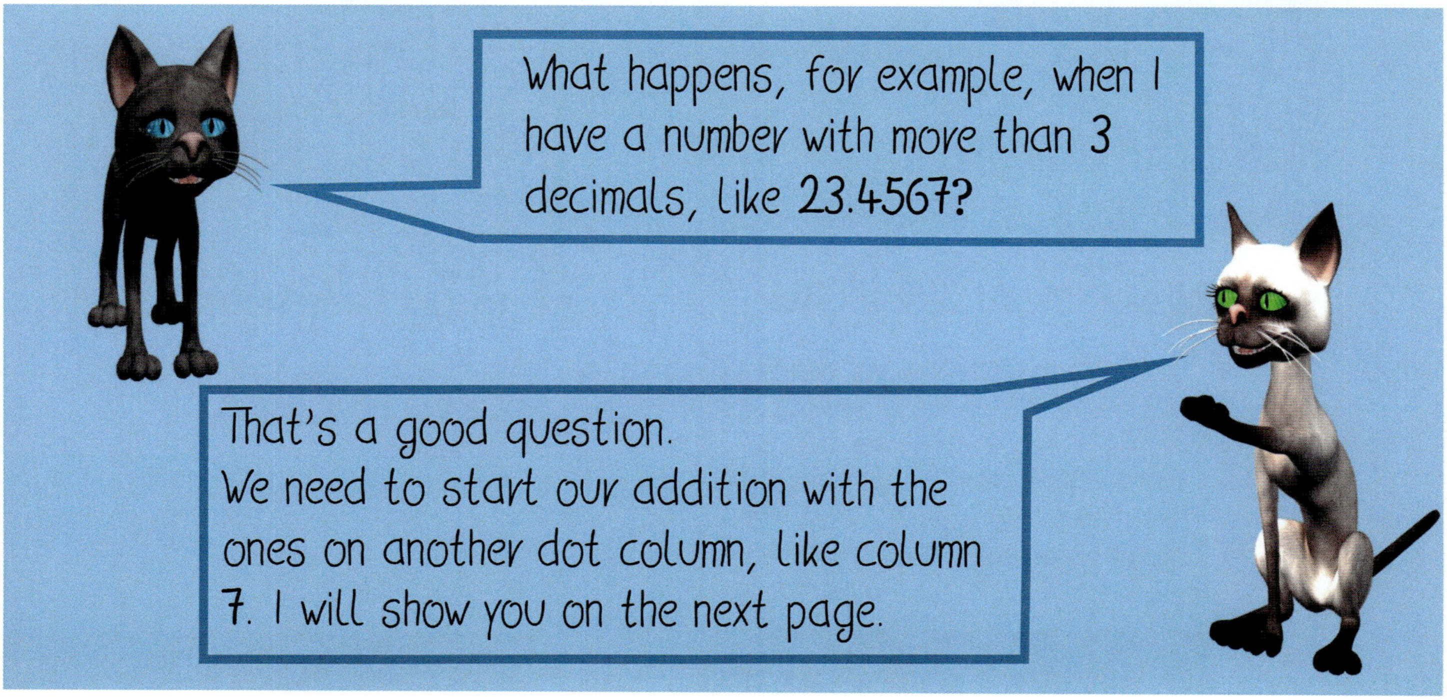

Adding decimal numbers with more than three decimals

We need to start by adding (or moving) a dot to a different place on the beam. For example we will move the dot to just before column 7. In this example 5.00235, the ones column will be on column 7.
So we have to put the 5 of 5.00235 on column 7.

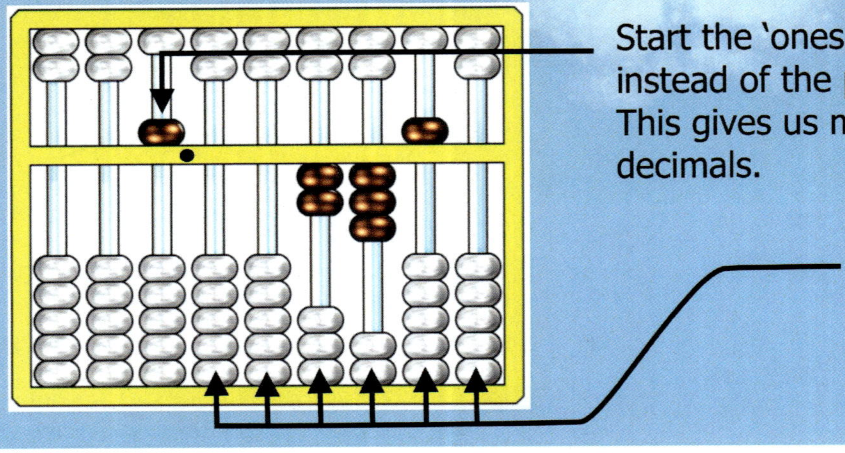

Start the 'ones column' here (column 7) instead of the previously used column 4. This gives us more columns for our decimals.

All of these columns can now be used for decimals.

Example: 200.00264 + 0.0012

Column 7 is our ones column

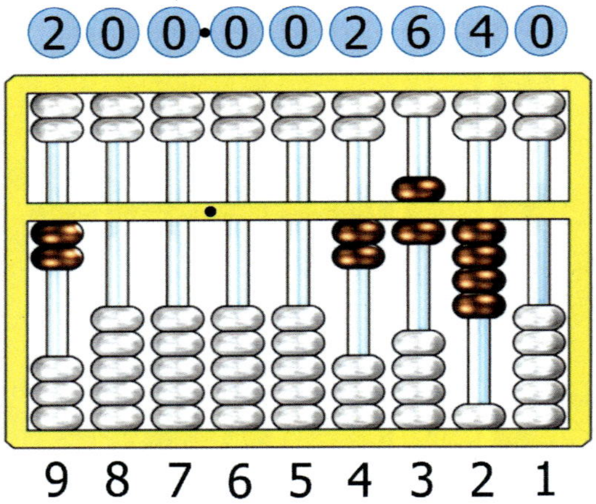

We will **register 200.00264**

200	• Column 9, register 2 lower beads
	• Columns 8, 7, 6 & 5, do nothing
0.002	• Column 4, register 2 lower beads
0.0006	• Column 3, register 1 upper and 1 lower bead
0.00004	• Column 4, register 4 lower beads

The abacus reads 200.00264

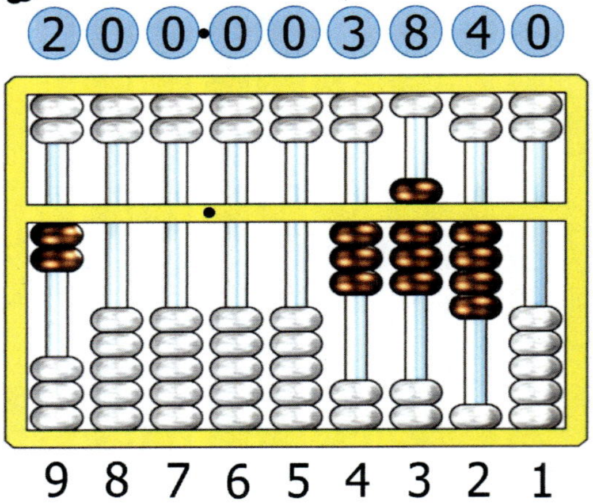

We will now **add 0.0012**

+0.001	• Column 4, register 1 lower bead
+0.0002	• Column 3, register 2 lower beads

The abacus result is 200.00384

TEST 3 - Addition of decimal numbers

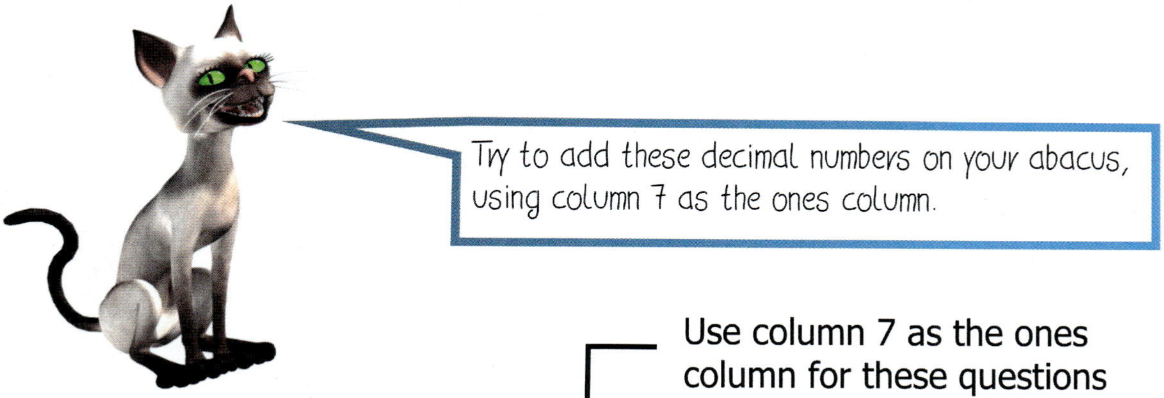

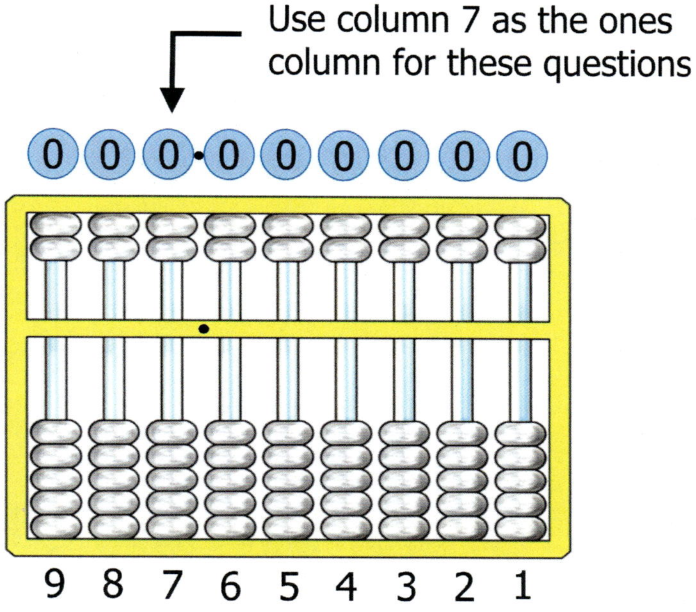

Answers are on page 48.

1. 43.0001 + 5.00004 =

2. 800.5239 + 0.1221

3. 0.2316 + 0.1021

4. 196.0246 + 4.2251

5. 272.1401 + 17.23002

Answers to test 3 (on page 47)

Here are the addition of decimal numbers answers.
If you got any wrong, just have another go.

1

43.0001 + 5.00004 = **48.00014**

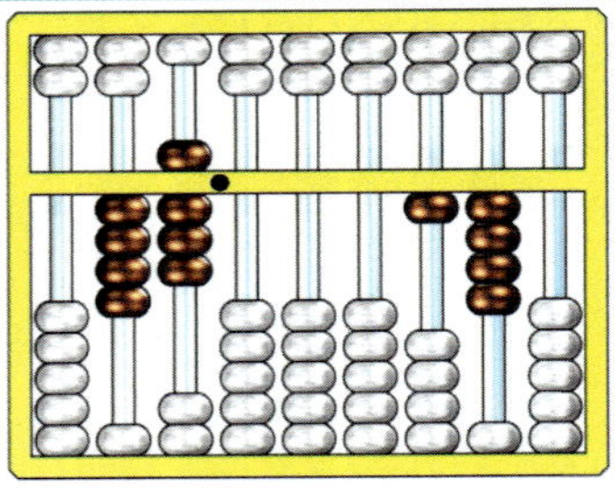

2

800.5239 + 0.1221 = **800.646**

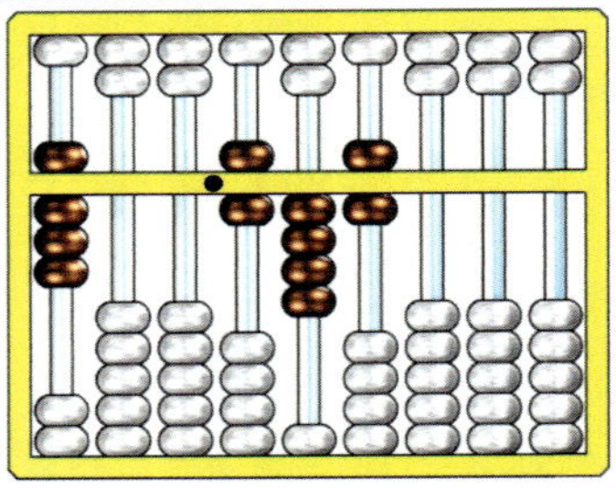

3

0.2316 + 0.1021 = **0.3337**

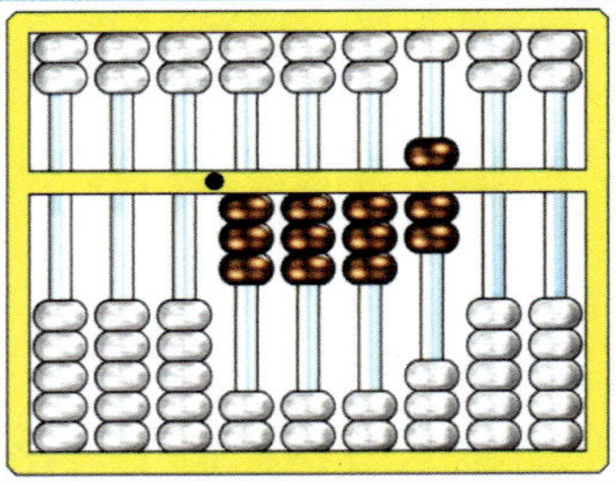

4

196.0246 + 4.2251 = **200.2497**

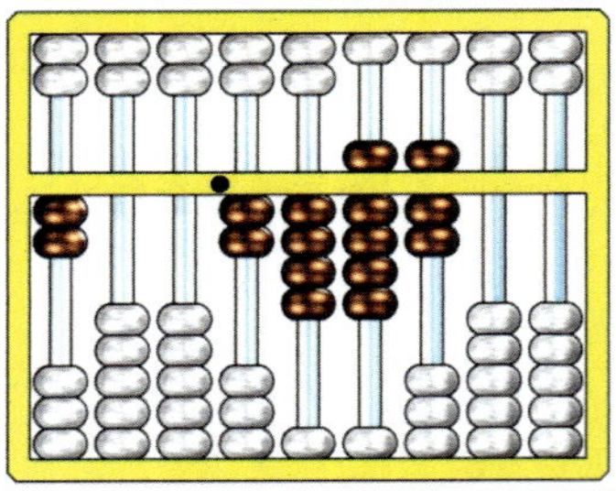

5

272.1401 + 17.23002 = **289.37012**

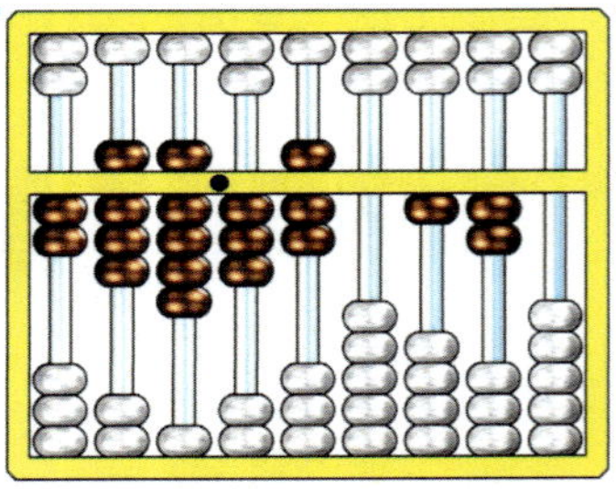

SUBTRACTION

Subtraction is taking one number away from another to find the difference.

Subtraction - things to remember:

- Register your numbers from left to right, just the same as we did with addition, for example: for number 231 register the 2 first, 3 second and 1 last.
- Each digit must be registered in the correct column, for example with 231 the 2 is for column 3 (hundredths column), the 3 for column 2 (tens column) and the 1 for column 1 (ones column), just like we did with addition.

Example: 432 - 221

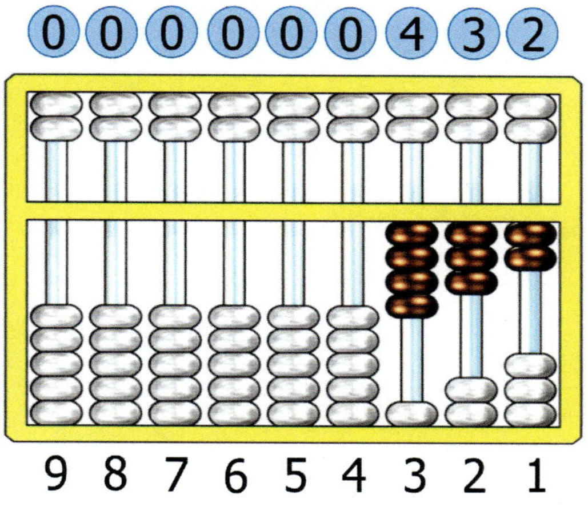

We will **register 432**

400	• Column 3, register 4 lower beads
30	• Column 2, register 3 lower beads
2	• Column 1, register 2 lower beads

The abacus reads 432

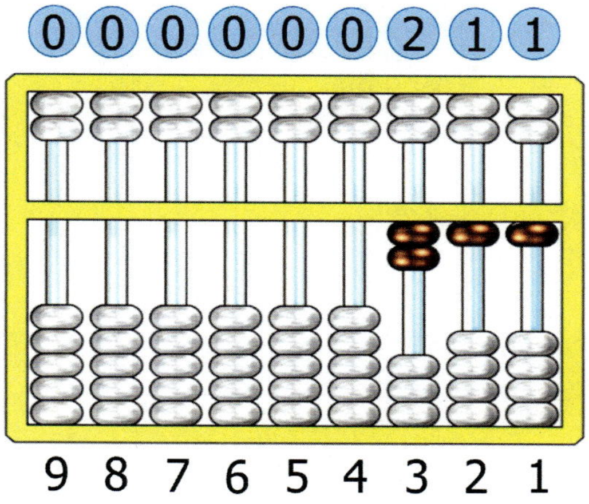

We will now **subtract 221** from 432

-200	• Column 3, unregister 2 lower beads to subtract 200
-20	• Column 2, unregister 2 lower beads to subtract 20
-1	• Column 1, unregister 1 lower bead to subtract 1

The abacus result is 211

These columns are useful to see the amount that you are subtracting.
For example:
30 means that you have just registered 30
-200 means that you have just subtracted 200

Subtracting numbers that have different amounts of digits

For example, when subtracting **4234 - 21** we see that 4234 has 4 digits and 21 only has 2.

Register the number that has the **largest amount of digits**, in this case it is **4234**.

Next, subtract the number with the smallest amount of digits, in this example the **21**, from the largest digit number.

More subtraction examples

Example: 4234 - 21

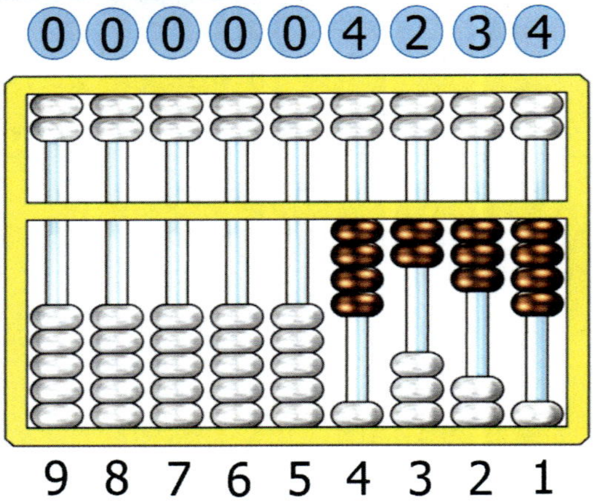

We will **register 4234**

4000 • Column 4, register 4 lower beads
200 • Column 3, register 2 lower beads
30 • Column 2, register 3 lower beads
4 • Column 1, register 4 lower beads

Register the number with the most digits first (4234)!

The abacus reads 4234

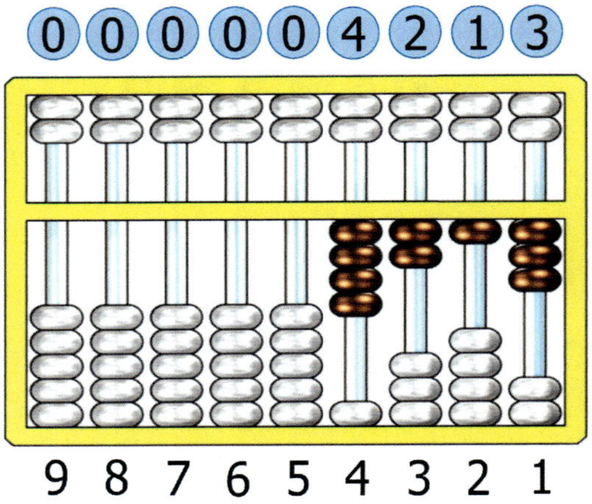

We will now **subtract 21**

-20 • Column 2, unregister 2 lower beads to subtract 20
-1 • Column 1, unregister 1 lower bead to subtract 1

The abacus result is 4213

More subtraction examples

Example: 9 - 5

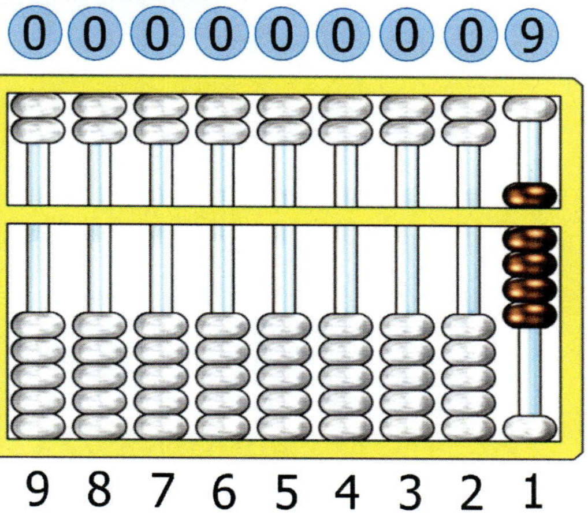

We will **register 9**

9 • Column 1, register 1 upper bead and 4 lower beads

The abacus reads 9

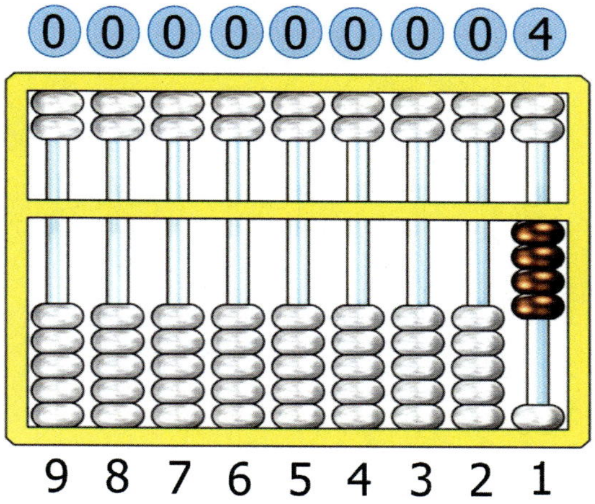

We will now **subtract 5**

-5 • Column 1, unregister 1 upper bead

The abacus result is 4

Example: 87 - 21

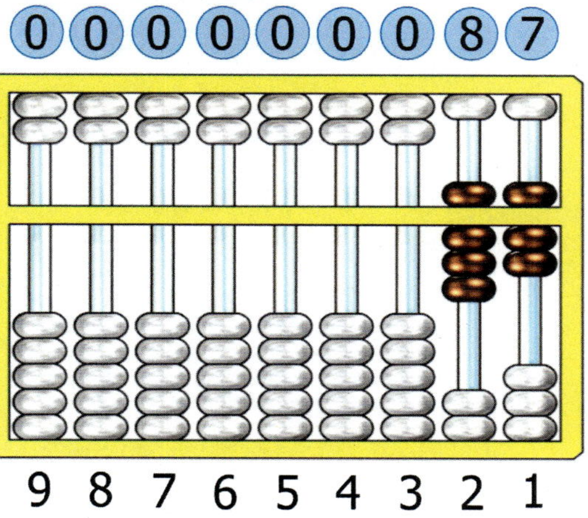

We will **register 87**

80 • Column 2, register 1 upper bead and 3 lower beads

7 • Column 1, register 1 upper bead and 2 lower beads

The abacus reads 87

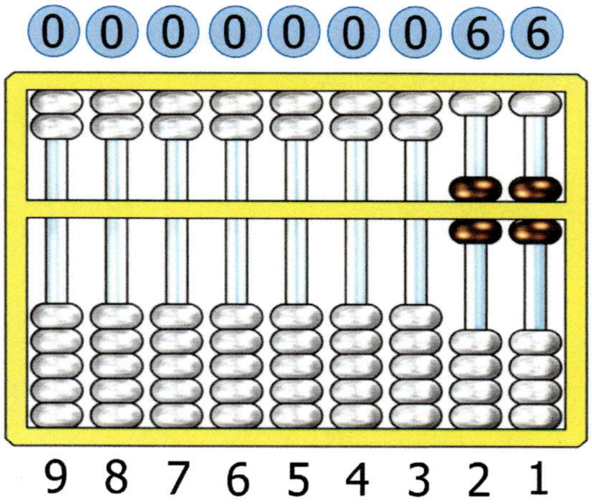

We will **subtract 21**

-20 • Column 2, unregister 2 lower beads

-1 • Column 1, unregister 1 lower bead

The abacus result is 66

More subtraction examples

Example: 987 - 671

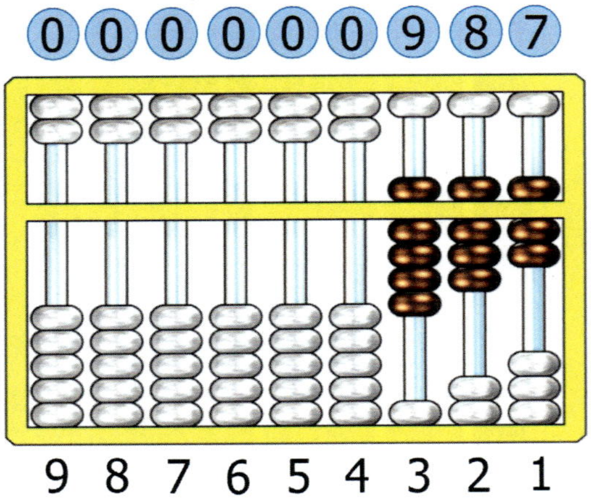

We will **register 987**

900 • Column 3, register 1 upper bead and 4 lower beads

80 • Column 2, register 1 upper bead and 3 lower beads

7 • Column 1, register 1 upper bead and 2 lower beads

The abacus reads 987

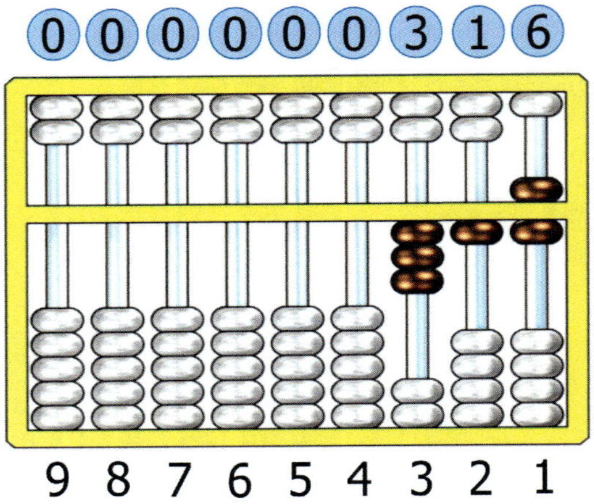

We will now **subtract 671**

-600 • Column 3, unregister 1 upper bead and 1 lower bead (Total = -500-100=-600)

-70 • Column 2, unregister 1 upper bead and 2 lower beads

-1 • Column 1, unregister 1 lower bead

The abacus result is 316

Example: 6533 - 323

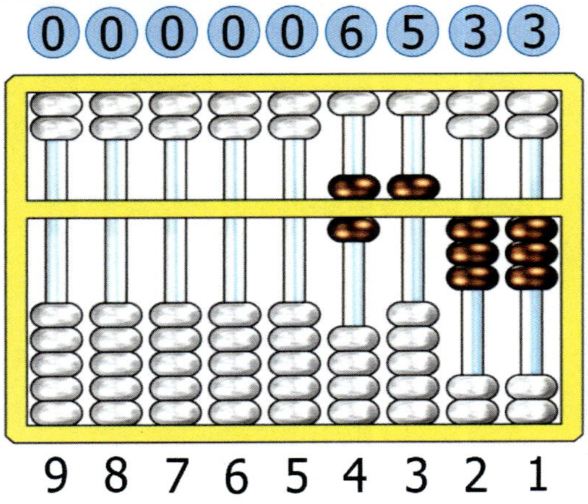

We will **register 6533**

6000 • Column 4, register 1 upper bead and 1 lower bead

500 • Column 3, register 1 upper bead

30 • Column 2, register 3 lower beads

3 • Column 1, register 3 lower beads

The abacus reads 6533

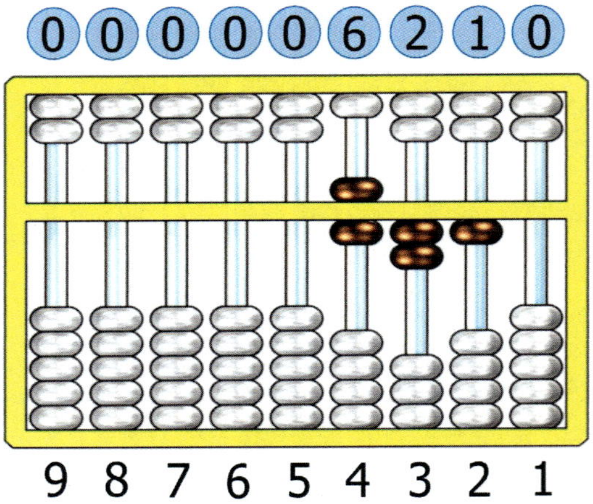

We will now **subtract 323**

-300 • Column 3, unregister 1 upper bead and register 2 lower beads (Total = -500+200=-300)

-20 • Column 2, unregister 2 lower beads

-3 • Column 1, unregister 3 lower beads

The abacus result is 6210

More subtraction examples

Example: 76205 - 5203

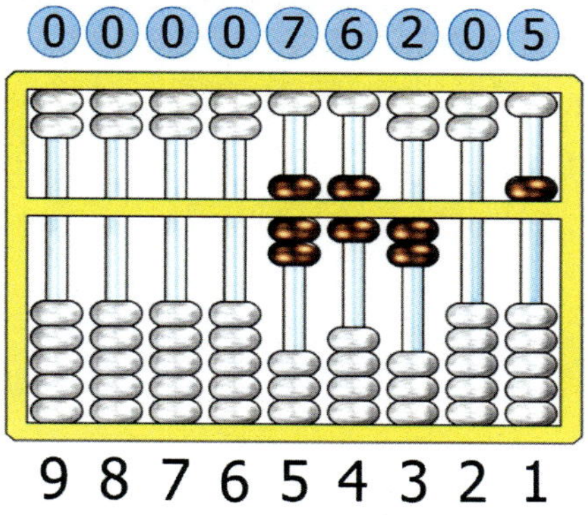

We will **register 76205**

- 70000 • Column 5, register 1 upper bead and 2 lower beads
- 6000 • Column 4, register 1 upper bead and 1 lower bead
- 200 • Column 3, register 2 lower beads
- • Column 2, do nothing
- 5 • Column 1, register 1 upper bead

The abacus reads 76205

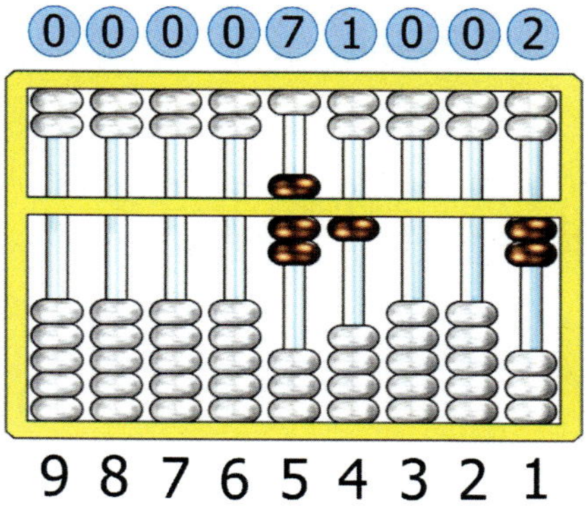

We will now **subtract 5203**

- -5000 • Column 4, unregister 1 upper bead
- -200 • Column 3, unregister 2 lower beads
- • Column 2, do nothing
- -3 • Column 1, unregister 1 upper bead and register 2 lower beads

The abacus result is 71002

Example: 642603284 - 521502064

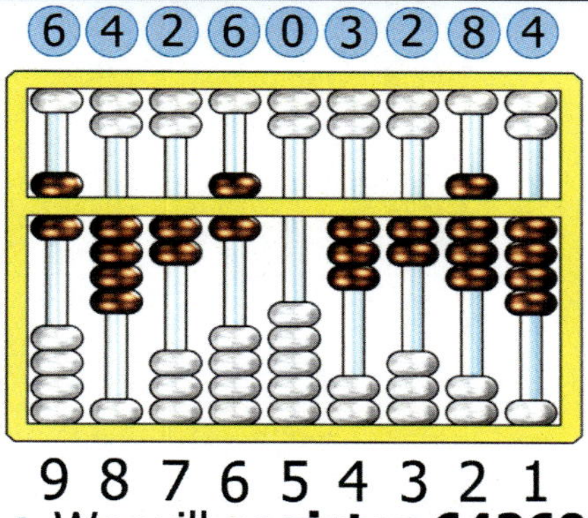

We will **register 642603284**

- 600000000 • Column 9, register 1 upper and 1 lower bead
- 40000000 • Column 8, register 4 lower beads
- 2000000 • Column 7, register 2 lower beads
- 600000 • Column 6, register 1 upper and 1 lower bead
- • Column 5, do nothing
- 3000 • Column 4, register 3 lower beads
- 200 • Column 3, register 2 lower beads
- 80 • Column 2, register 1 upper and 3 lower beads
- 4 • Column 1, register 4 lower beads

The abacus reads 642603284

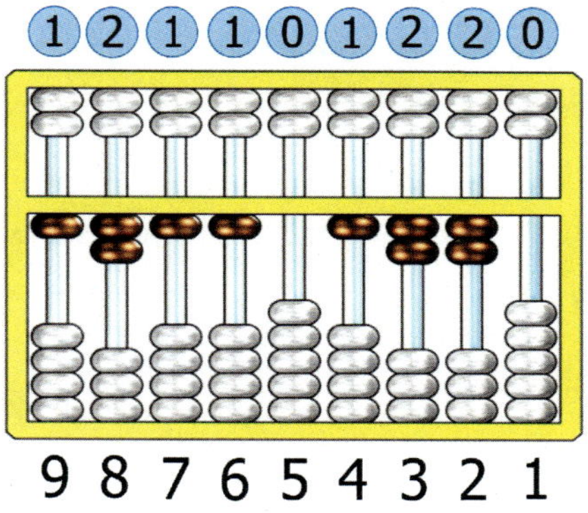

We will **subtract 521502064**

- -500000000 • Column 9, unregister 1 upper bead
- -20000000 • Column 8, unregister 2 lower beads
- -1000000 • Column 7, unregister 1 lower bead
- -500000 • Column 6, unregister 1 upper bead
- • Column 5, do nothing
- -2000 • Column 4, unregister 2 lower beads
- • Column 3, do nothing
- -60 • Column 2, unregister 1 upper and 1 lower bead
- -4 • Column 1, unregister 4 lower beads

The abacus result is 121101220

When there are not enough beads in the column for the subtraction

When you don't have enough beads, move to the next LEFT column to help

For example, when you try to subtract 8 from an already registered number 12, you don't have enough beads in the column where the 2 of the 12 is, to do it. You can only unregister a maximum of 9 in each column (4 lower beads and 1 upper bead, -4-5=-9).
When this happens, we need to use the
'Not enough beads list for subtraction'.

-1=-10+9
-2=-10+8
-3=-10+7
-4=-10+6
-5=-10+5
-6=-10+4
-7=-10+3
-8=-10+2
-9=-10+1

How to use the 'Not enough beads list for subtraction'

Let's say we need to subtract 8 from a column but we don't have enough beads.

Look at the list, **-8=-10+2**

10 is the number to **unregister,** in the next **LEFT** column (1 lower bead).

2 is the number to **register** in our column**.**

Example: 12 - 8

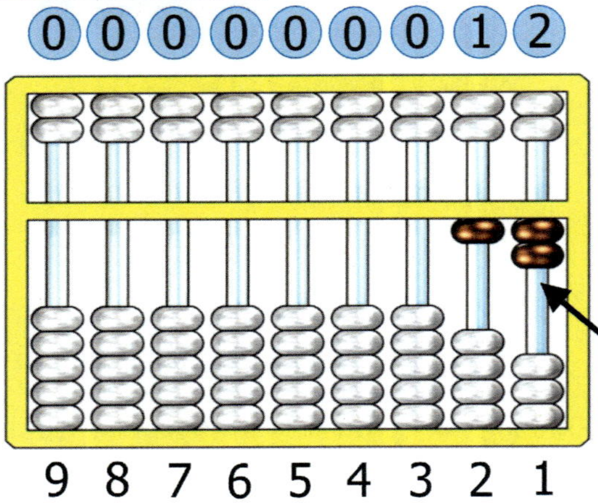

We will **register 12**

10 • Column 2, register 1 lower bead
2 • Column 1, register 2 lower beads

Unregister means move away from the beam (subtract)!

The abacus reads 12

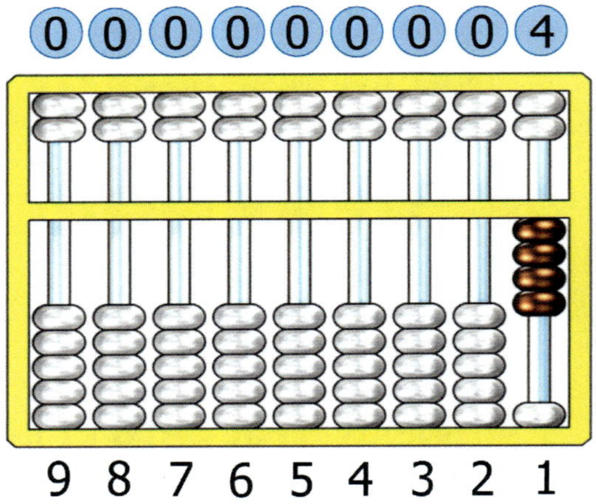

We will now **subtract 8**

There are not enough beads in column 1 to subtract 8, move to the next LEFT column to help. First we must think **-8=-10+2** see the **'Not enough beads list for subtraction'**

-10 • Column 2, **unregister** 1 lower bead to subtract 10
+2 • Column 1, **register** 2 lower beads to add 2 (from -8=-10+2)

The abacus result is 4

More subtraction examples (when we don't have enough beads)

Example: 25 - 16

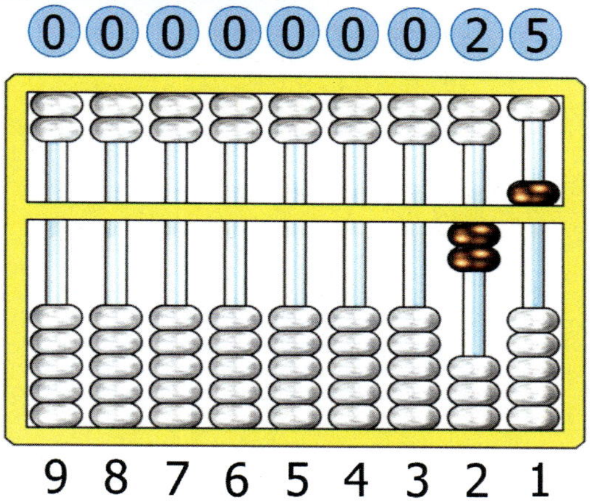

We will **register 25**

20 • Column 2, register 2 lower beads

5 • Column 1, register 1 upper bead

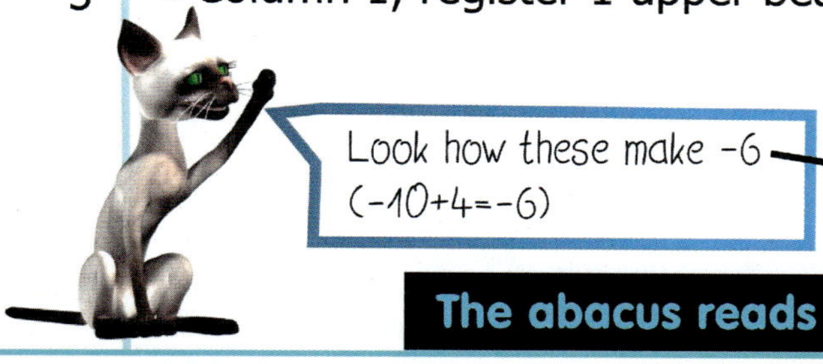

The abacus reads 25

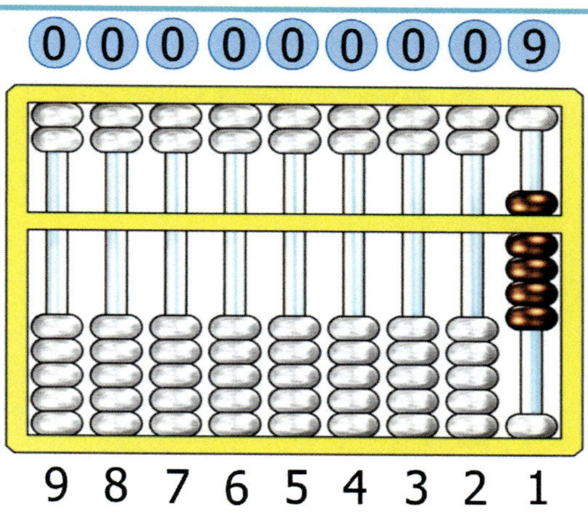

We will now **subtract 16**

-10 • Column 2, unregister 1 lower bead

There are not enough beads in column 1 to subtract 6, so move to the next left column to help and think **-6=-10+4**

-10 • Column 2, unregister 1 lower bead to subtract 10

+4 • Column 1, register 4 lower beads to add 4

The abacus result is 9

Example: 10 - 5

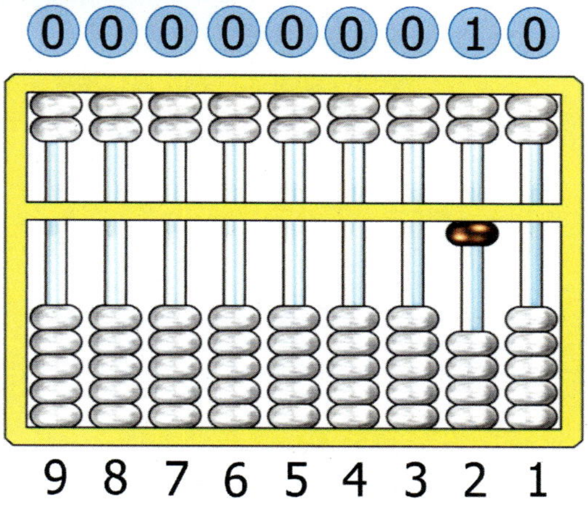

We will **register 10**

10 • Column 2, register 1 lower bead

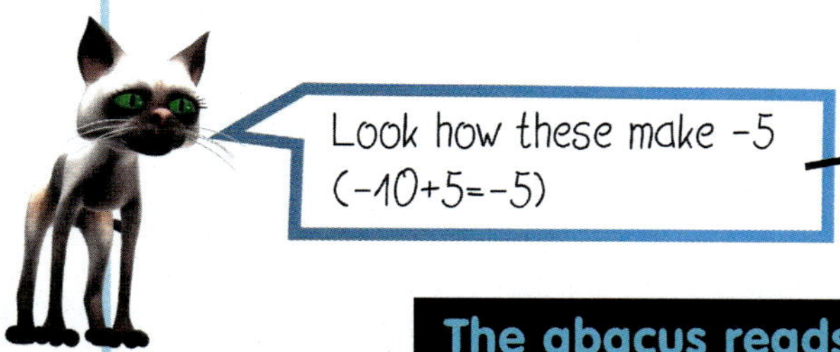

The abacus reads 10

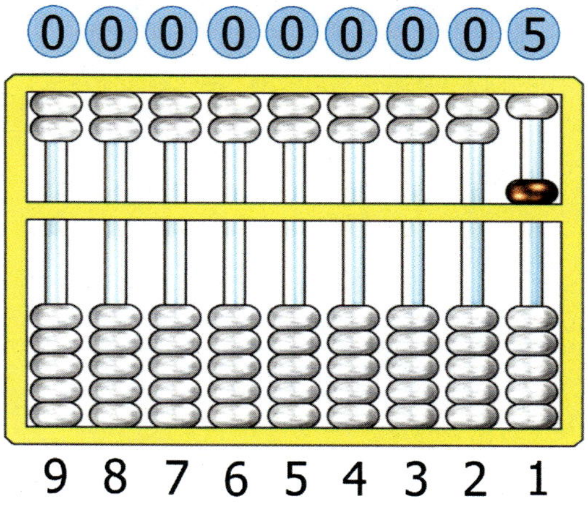

We will **subtract 5**

There are not enough beads in column 1 to subtract 5, so move to the next left column to help and think **-5=-10+5**

-10 • Column 2, unregister 1 lower bead

+5 • Column 1, register 1 upper bead

The abacus result is 5

More subtraction examples (when we don't have enough beads)

Example: 477 - 286

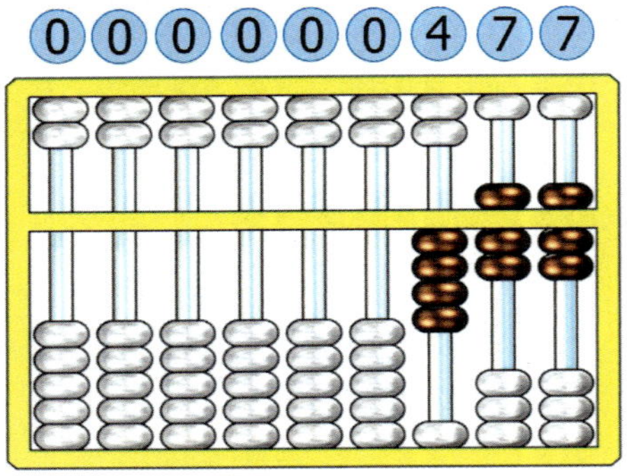

9 8 7 6 5 4 3 2 1

We will **register 477**

400 • Column 3, register 4 lower beads
70 • Column 2, register 1 upper bead and 2 lower beads
7 • Column 1, register 1 upper bead and 2 lower beads

The abacus reads 477

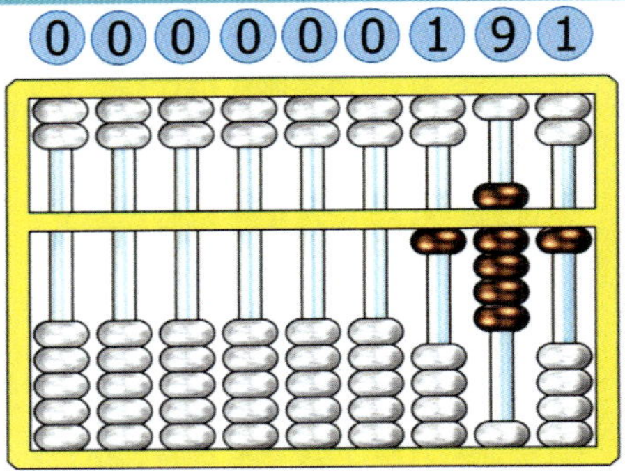

9 8 7 6 5 4 3 2 1

We will now **subtract 286**

-200 • Column 3, unregister 2 lower beads

There are not enough beads in column 2 to subtract 8, so move to the next left column to help and think **-8=-10+2**

-100 • Column 3, unregister 1 lower bead
+20 • Column 2, register 2 lower beads
-6 • Column 1, unregister 1 upper and 1 lower bead

The abacus result is 191

Example: 463 - 386

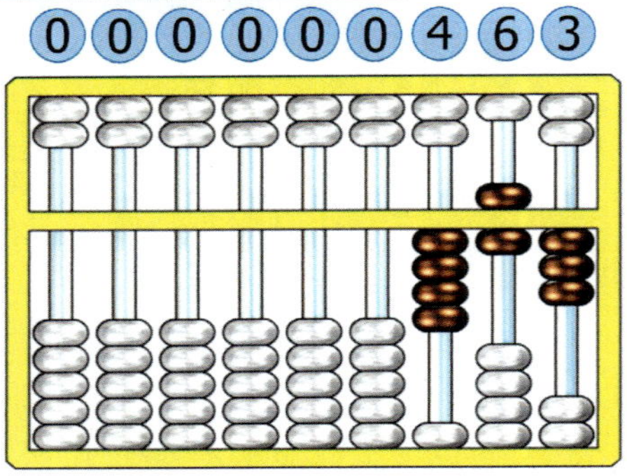

9 8 7 6 5 4 3 2 1

We will **register 463**

400 • Column 3, register 4 lower beads
60 • Column 2, register 1 upper bead and 1 lower bead
3 • Column 1, register 3 lower beads

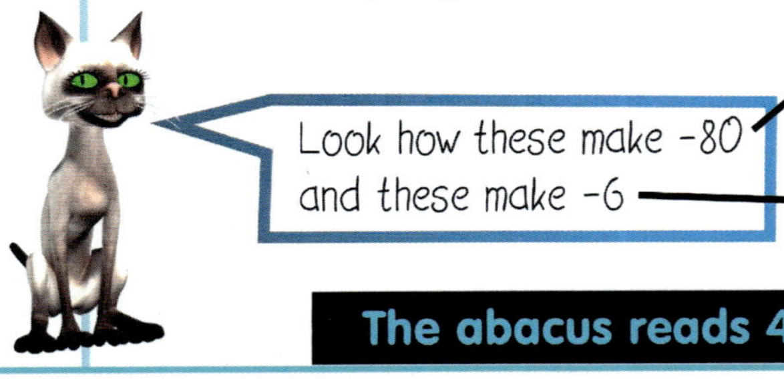

The abacus reads 463

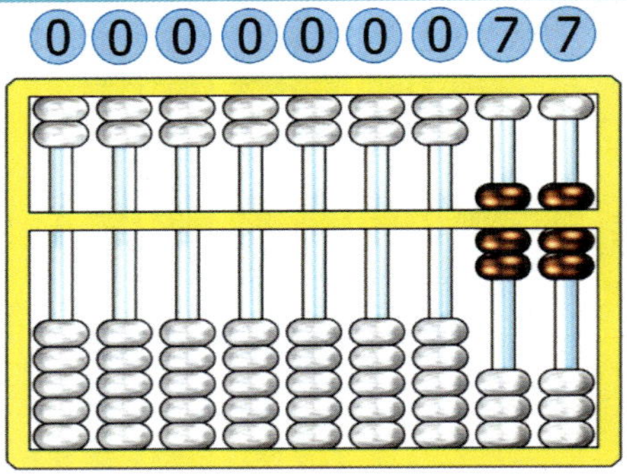

9 8 7 6 5 4 3 2 1

We will **subtract 386**

-300 • Column 3, unregister 3 lower beads

There are not enough beads in column 2 to subtract 8, so move to the next left column to help and think **-8=-10+2**

-100 • Column 3, unregister 1 lower bead
+20 • Column 2, register 2 lower beads

There are not enough beads in column 1 to subtract 6, so move to the next left column to help and think **-6=-10+4**

-10 • Column 2, unregister 1 lower bead
+4 • Column 1, register 1 upper bead and unregister 1 lower bead (+5-1=+4)

The abacus result is 77

More subtraction examples (when we don't have enough beads)

Example: 6533 - 600

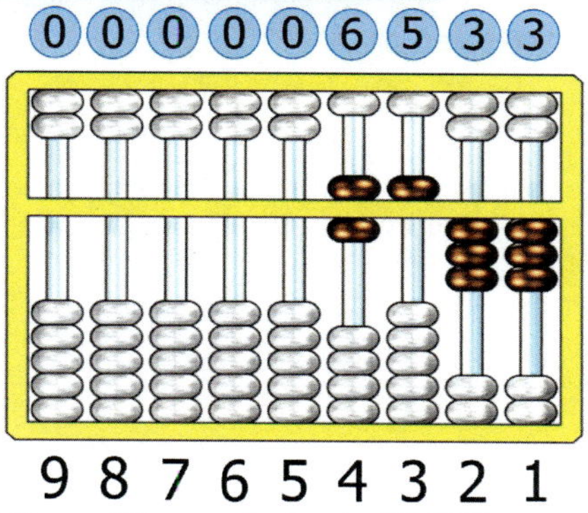

We will **register 6533**

- 6000 • Column 4, register 1 upper and 1 lower bead
- 500 • Column 3, register 1 upper bead
- 30 • Column 2, register 3 lower beads
- 3 • Column 1, register 3 lower beads

The abacus reads 6533

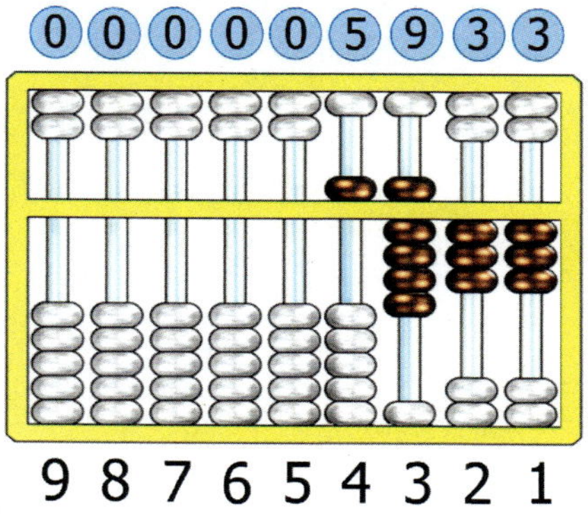

We will now **subtract 600**

There are not enough beads in column 3 to subtract 6, so move to the next left column to help and think **-6=-10+4**

- -1000 • Column 4, unregister 1 lower bead
- +400 • Column 3, register 4 lower beads
- • Columns 2 & 1, do nothing

The abacus result is 5933

Example: 6495613 - 5530232

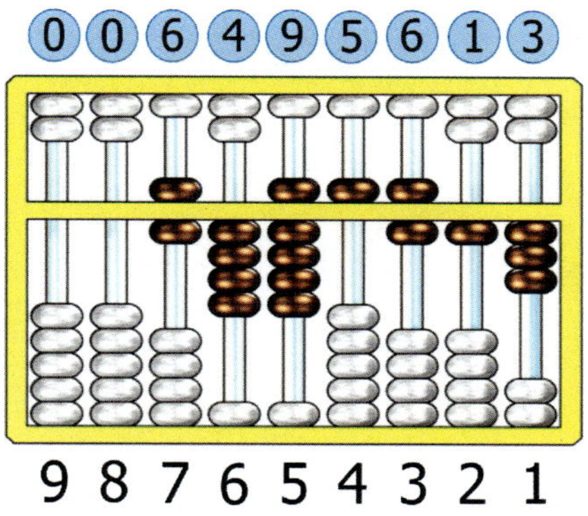

We will **register 6495613**

- 6000000 • Column 7, register 1 upper and 1 lower bead
- 400000 • Column 6, register 4 lower beads
- 90000 • Column 5, register 1 upper and 4 lower beads
- 5000 • Column 4, register 1 upper bead
- 600 • Column 3, register 1 upper and 1 lower bead
- 10 • Column 2, register 1 lower bead
- 3 • Column 1, register 3 lower beads

The abacus reads 6495613

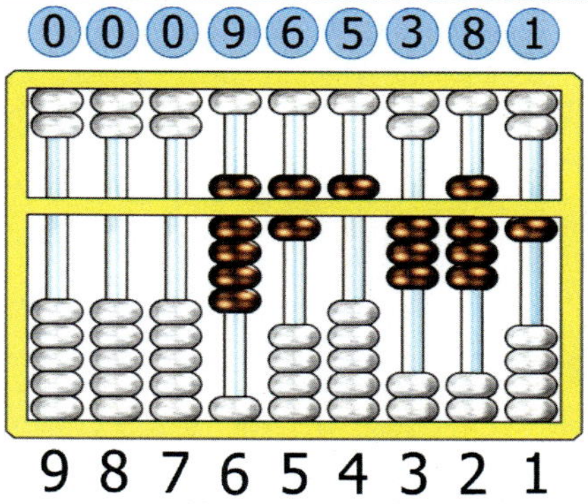

We will now **subtract 5530232**

- -5000000 • Column 7, unregister 1 upper bead

Not enough beads in column 6 to subtract 5, so think **-5=-10+5**

- -1000000 • Column 7, unregister 1 lower bead
- +500000 • Column 6, register 1 upper bead
- -30000 • Column 5, unregister 3 lower beads
- • Column 4, do nothing
- -200 • Column 3, unregister 1 upper bead and register 3 lower beads

Not enough beads in column 2 to subtract 3, so think **-3=-10+7**

- -100 • Column 3, unregister 1 lower bead
- +50 • Column 2, register 1 upper bead
- +20 • Column 2, register 2 lower beads
- -2 • Column 1, unregister 2 lower beads

The abacus result is 965381

More subtraction examples (when we don't have enough beads)

Example: 100000 - 11111

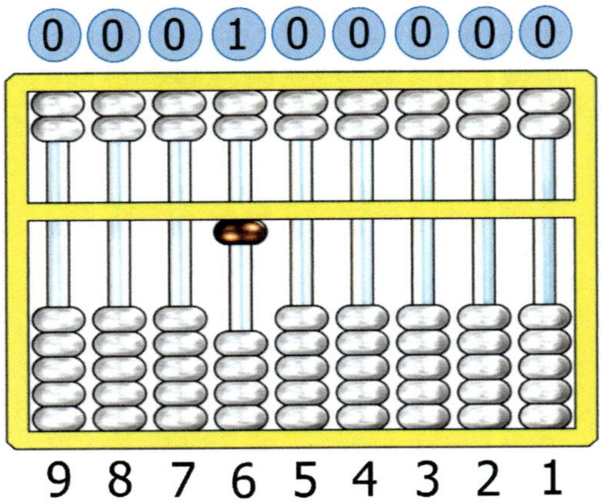

We will **register 100000**

100000 • Column 6, register 1 lower bead

Look what these make:
-10000
-1000
-100
-10
-1

The abacus reads 100000

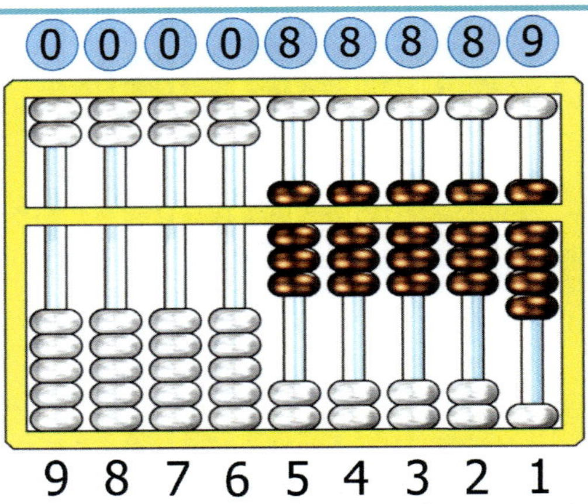

We will now **subtract 11111**

-100000
+90000

Not enough beads in column 5 to subtract 1, move to column 6 and think **-1=-10+9**

- Column 6, unregister 1 lower bead
- Column 5, register 1 upper and 4 lower beads

-10000
+9000

Not enough beads in column 4 to subtract 1, move to column 5 and think **-1=-10+9**

- Column 5, unregister 1 lower bead
- Column 4, register 1 upper and 4 lower beads

-1000
+900

Not enough beads in column 3 to subtract 1, move to column 4 and think **-1=-10+9**

- Column 4, unregister 1 lower bead
- Column 3, register 1 upper and 4 lower beads

-100
+90

Not enough beads in column 2 to subtract 1, move to column 3 and think **-1=-10+9**

- Column 3, unregister 1 lower bead
- Column 2, register 1 upper and 4 lower beads

-10
+9

Not enough beads in column 1 to subtract 1, move to column 2 and think **-1=-10+9**

- Column 2, unregister 1 lower bead
- Column 1, register 1 upper and 4 lower beads

The abacus result is 88889

Skipped columns when subtracting

Like we did with addition, sometimes with subtraction we have to SKIP a column. I'll explain why below.

When a column doesn't have enough beads left on it to make the subtraction, we move to the next LEFT column to help. Sometimes the next left column also doesn't have enough beads on it, so we **SKIP** this column and move again to the next left column until you reach a column that has enough beads to use. See below how it works.

Important!

- We will **SKIP** a column when there are not enough beads to use in that column.
- We will see this symbol when we need to **SKIP** ← a column (move on to the next left column).
- We will see this symbol when we need to **MOVE BACK** → a column (move on to the next right column).
- We will **REGISTER** all beads in any skipped columns (with addition we unregistered, here we do the opposite).

Example: 100 - 5

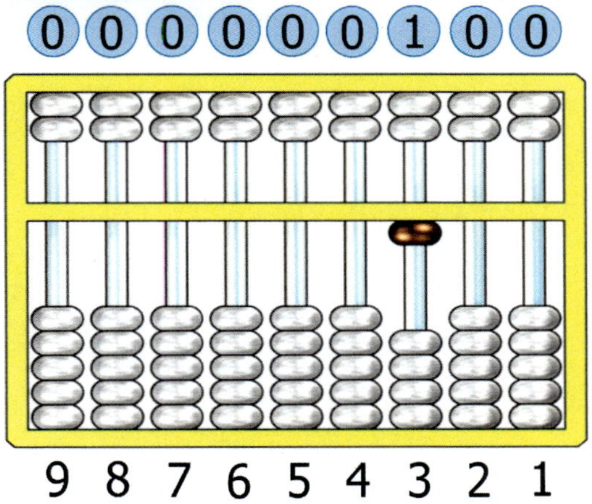

We will **register 100**

100 • Column 3, register 1 lower bead

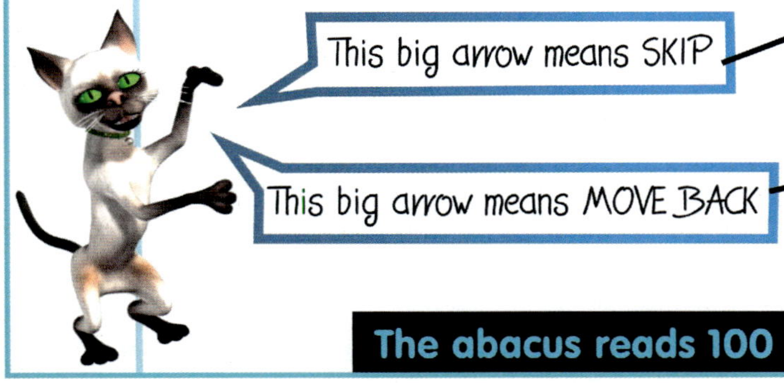

The abacus reads 100

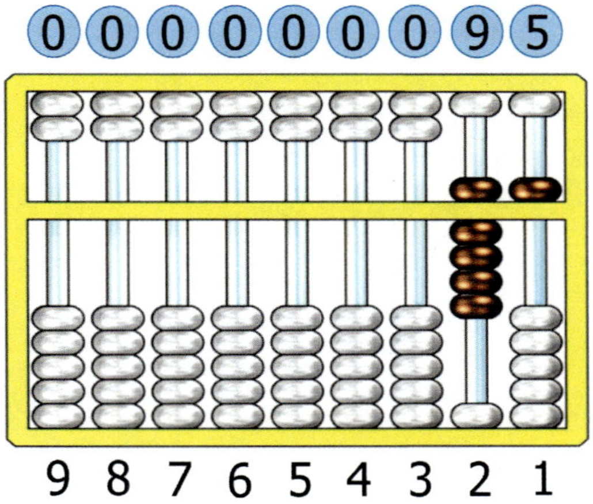

We will now **subtract 5**

There are not enough beads in column 1 to subtract 5, move to column 2, think **-5=-10+5**

← +90 • Column 2, **SKIP** this column and register all beads

-100 • Column 3, unregister 1 lower bead

→ • **MOVE BACK** past the skipped column 2

+5 • Column 1, register 1 upper bead

The abacus result is 95

More skipped column examples

Example: 1000 - 1

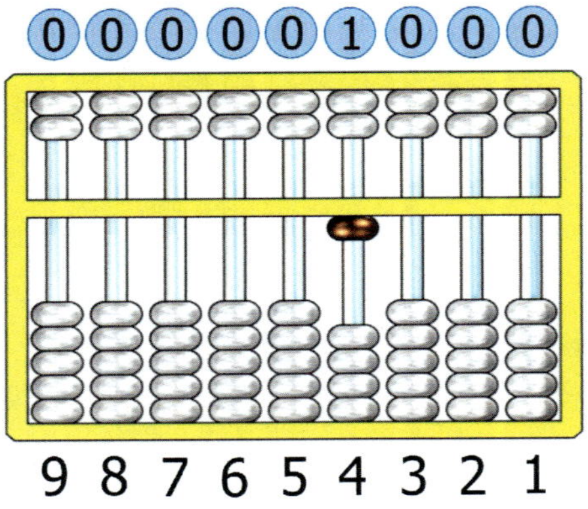

We will **register 1000**

1000 • Column 4, register 1 lower bead

> Remember to move back past the skipped columns!

The abacus reads 1000

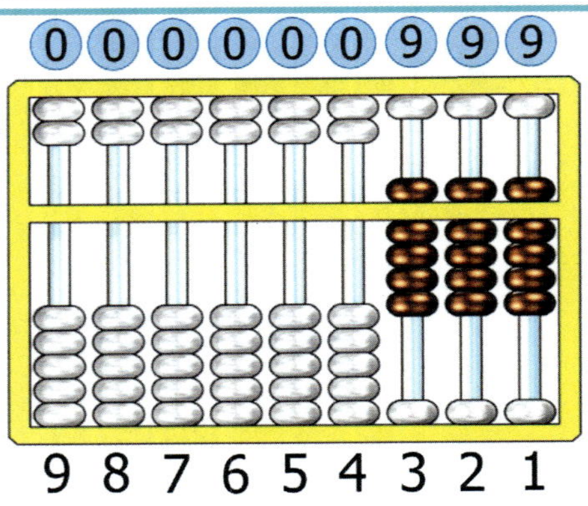

We will now **subtract 1**

There are not enough beads in column 1 to unregister 1, move to column 2, think **-1=-10+9**

- ← Column 2, **SKIP** this column and register all beads (+90)
- ← Column 3, **SKIP** this column and register all beads (+900)
- -1000 Column 4, unregister 1 lower bead
- → → **MOVE BACK** past the skipped columns 3 and 2
- +9 Column 1, register 1 upper and 4 lower beads

The abacus result is 999

Example: 204 - 5

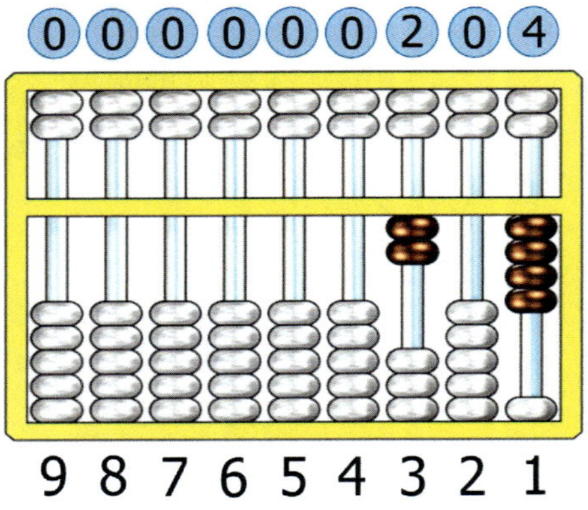

We will **register 204**

- 200 Column 3, register 2 lower beads
- Column 2, do nothing
- 4 Column 1, register 4 lower beads

The abacus reads 204

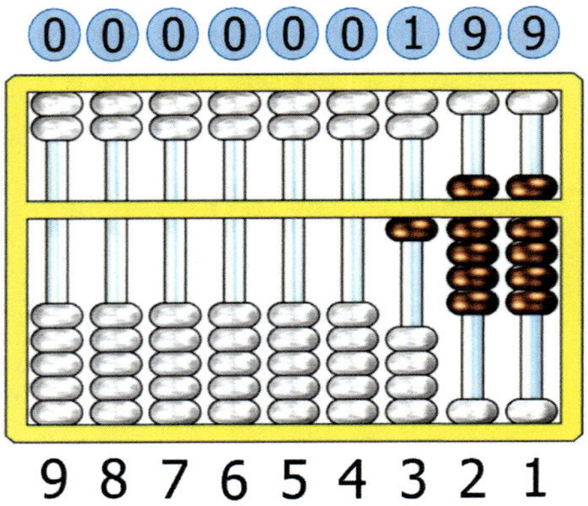

We will now **subtract 5**

There are not enough beads in column 1 to unregister 5, move to column 2, think **-5=-10+5**

- ← Column 2, **SKIP** this column and register all beads (+90)
- -100 Column 3, unregister 1 lower bead
- → **MOVE BACK** past the skipped column 2
- +5 Column 1, register 1 upper bead

The abacus result is 199

Subtraction of more than two numbers

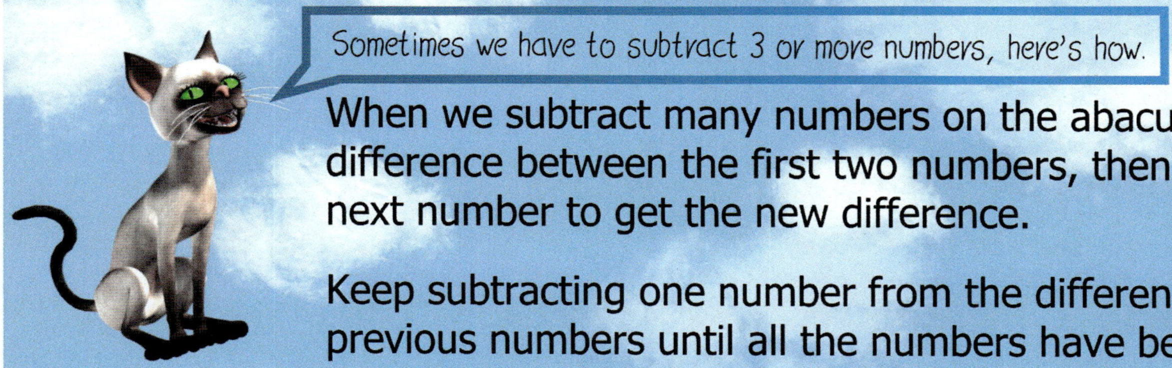

When we subtract many numbers on the abacus, just find the difference between the first two numbers, then subtract the next number to get the new difference.

Keep subtracting one number from the difference of the previous numbers until all the numbers have been subtracted.

Example: 998 - 332 - 151

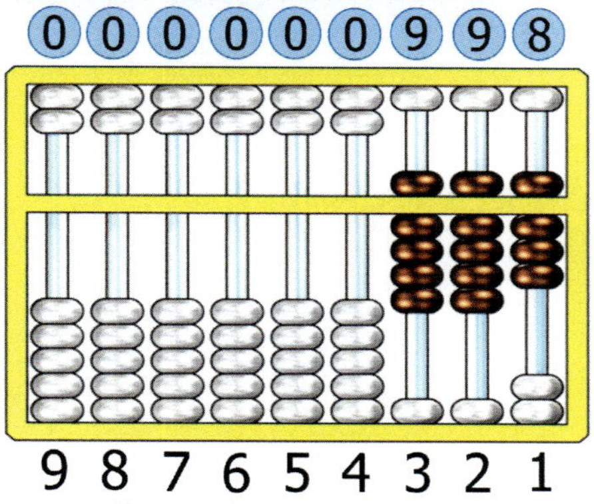

We will **register 998**

900 • Column 3, register 1 upper and 4 lower beads
90 • Column 2, register 1 upper and 4 lower beads
8 • Column 1, register 1 upper and 3 lower beads

The abacus reads 998

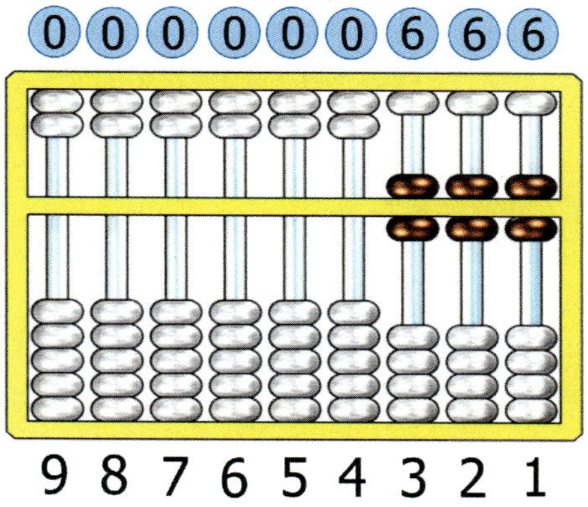

We will now **subtract 332 from 998**

-300 • Column 3, unregister 3 lower beads
-30 • Column 2, unregister 3 lower beads
-2 • Column 1, unregister 2 lower beads

The abacus now displays 666

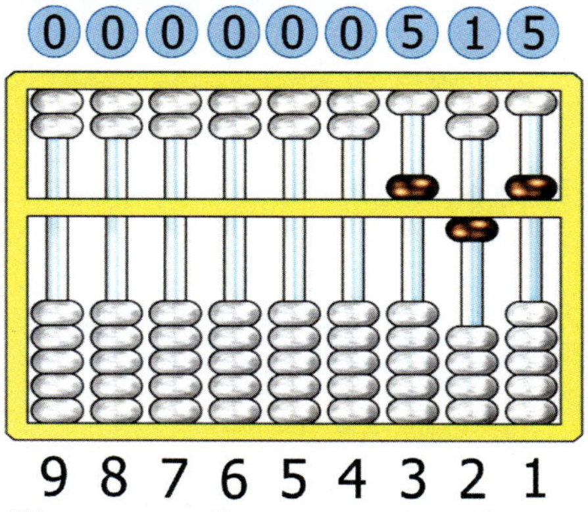

We will now **subtract 151 from 666**

-100 • Column 3, unregister 1 lower bead
-50 • Column 2, unregister 1 upper bead
-1 • Column 1, unregister 1 lower bead

The abacus result is 515

Subtraction of more than two numbers

Example: 244977999 - 121521210 - 123020163

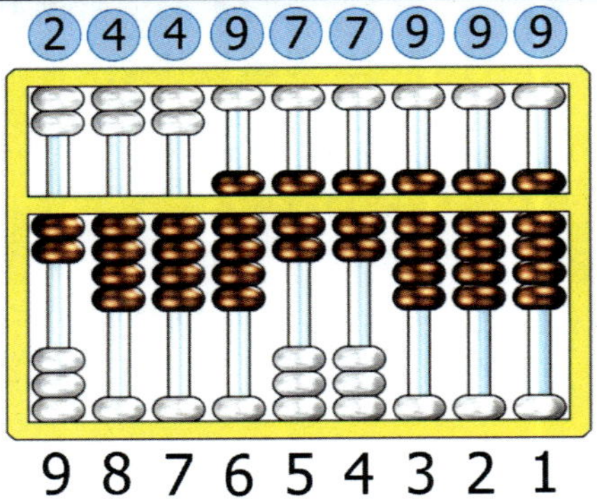

We will **register 244977999**

200000000	• Column 9, register 2 lower beads
40000000	• Column 8, register 4 lower beads
4000000	• Column 7, register 4 lower beads
900000	• Column 6, register 1 upper bead and 4 lower beads
70000	• Column 5, register 1 upper bead and 2 lower beads
7000	• Column 4, register 1 upper bead and 2 lower beads
900	• Column 3, register 1 upper bead and 4 lower beads
90	• Column 2, register 1 upper bead and 4 lower beads
9	• Column 1, register 1 upper bead and 4 lower beads

The abacus reads 43612424662

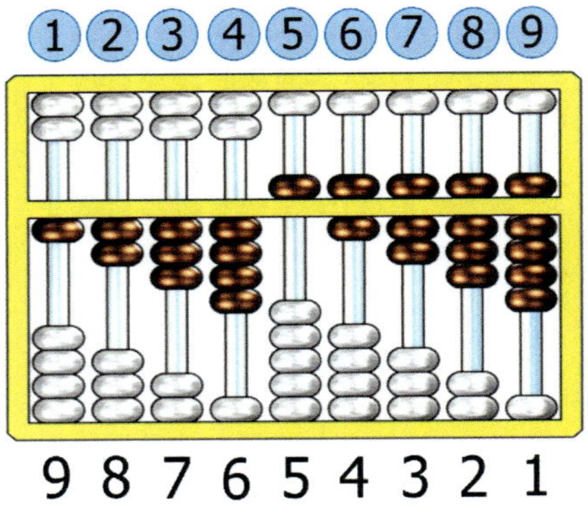

We will now **subtract 121521210 from 244977999**

-100000000	• Column 9, unregister 1 lower bead
-20000000	• Column 8, unregister 2 lower beads
-1000000	• Column 7, unregister 1 lower bead
-500000	• Column 6, unregister 1 upper bead
-20000	• Column 5, unregister 2 lower beads
-1000	• Column 4, unregister 1 lower bead
-200	• Column 3, unregister 2 lower beads
-10	• Column 2, unregister 1 lower bead
	• Column 1, do nothing

The abacus now displays 123456789

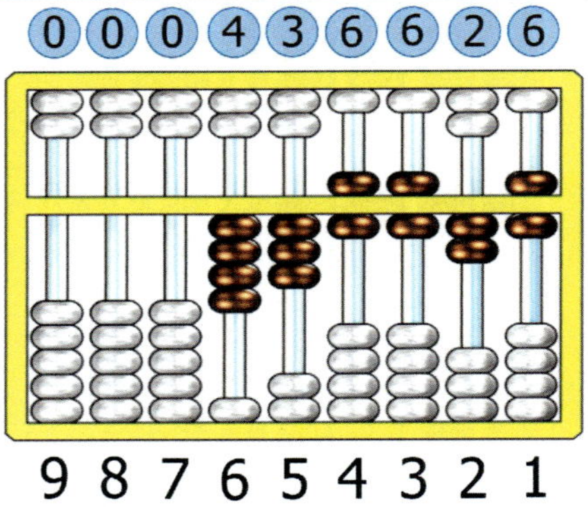

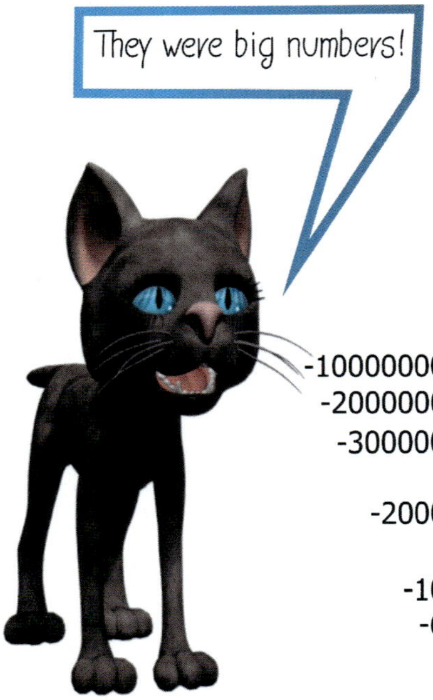

We will now **subtract 123020163 from 123456789**

-100000000	• Column 9, unregister 1 lower bead
-20000000	• Column 8, unregister 2 lower beads
-3000000	• Column 7, unregister 3 lower beads
	• Column 6, do nothing
-20000	• Column 5, unregister 2 lower beads
	• Column 4, do nothing
-100	• Column 3, unregister 1 lower bead
-60	• Column 2, unregister 1 upper and 1 lower bead
-3	• Column 1, unregister 3 lower beads

The abacus result is 436626

TEST 4 - Subtraction

Try to subtract these numbers on your abacus.

Answers are on pages 64, 65 and 66.

1. 92 - 35

2. 645 - 232

3. 814 - 8

4. 4342 - 2312

5. 66338 - 11223

6. 635253 - 246062

7. 5213634 - 6128

8. 4536 - 224 - 12

9. 440262 - 3201 - 435 - 6713

10. 87654321 - 10000008 - 642107

Answers to test 4 (on page 63)

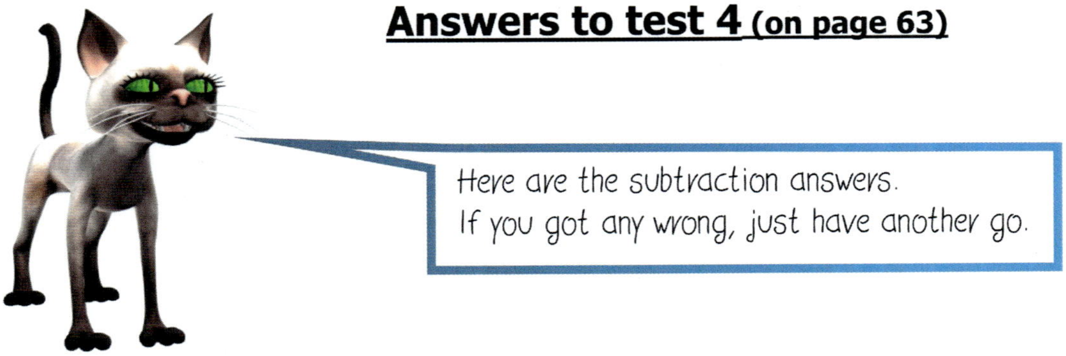

1

92 - 35 = **57**

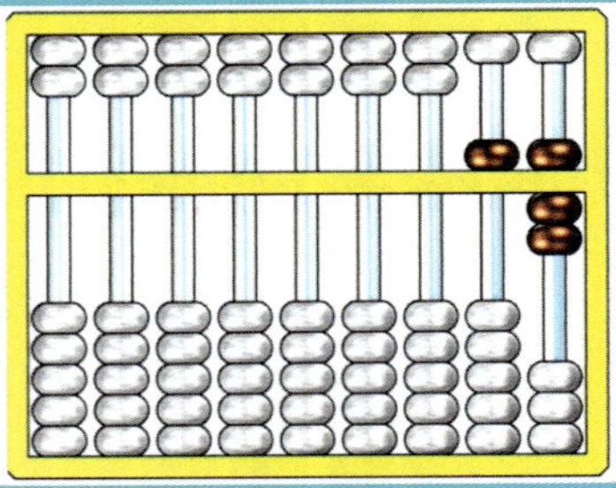

2

645 - 232 = **413**

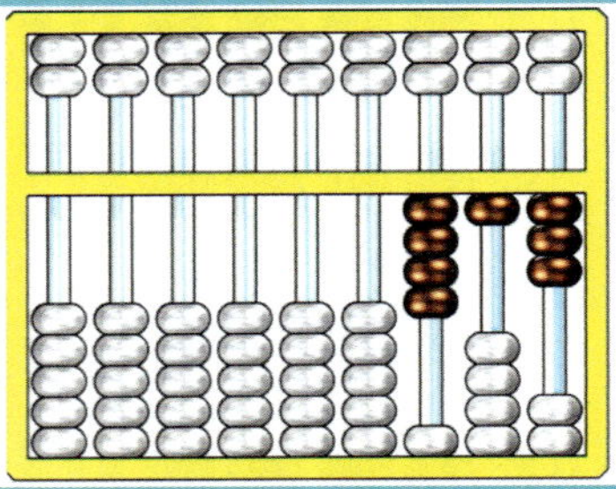

3

814 - 8 = **806**

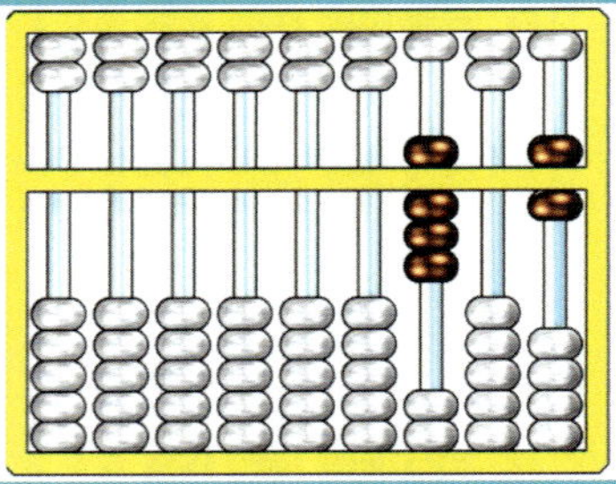

4

4342 - 2312 = **2030**

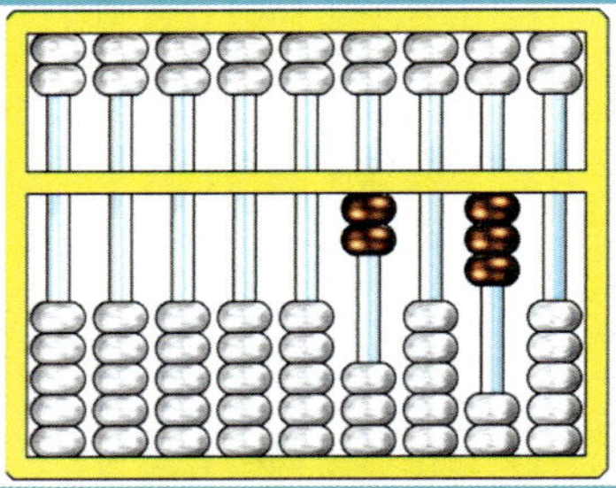

5

66338 - 11223 = **55115**

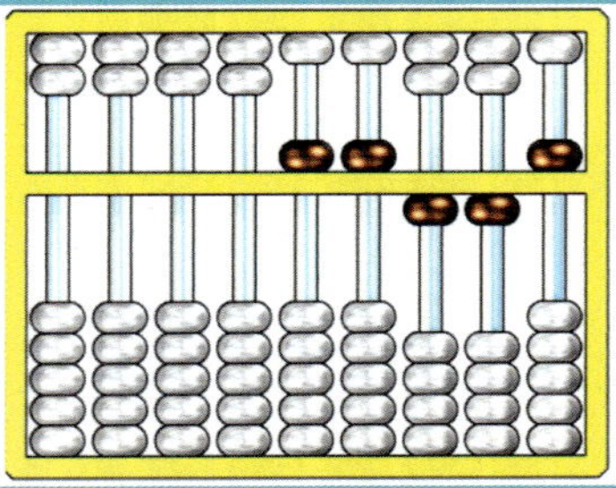

Answers to test 4 (on page 63)

6. 635253 - 246062 = **389191**

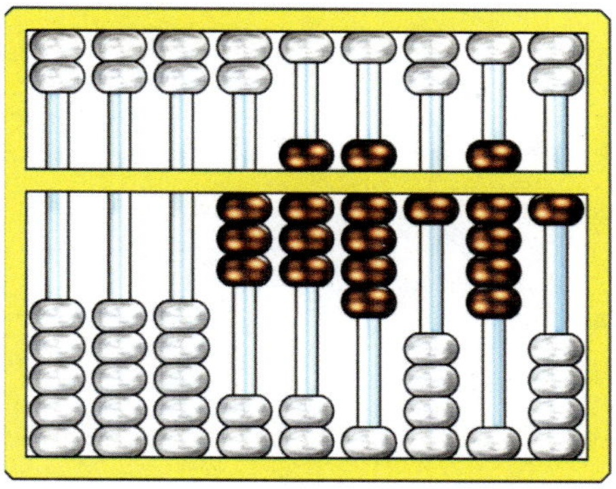

7. 5213634 - 6128 = **5207506**

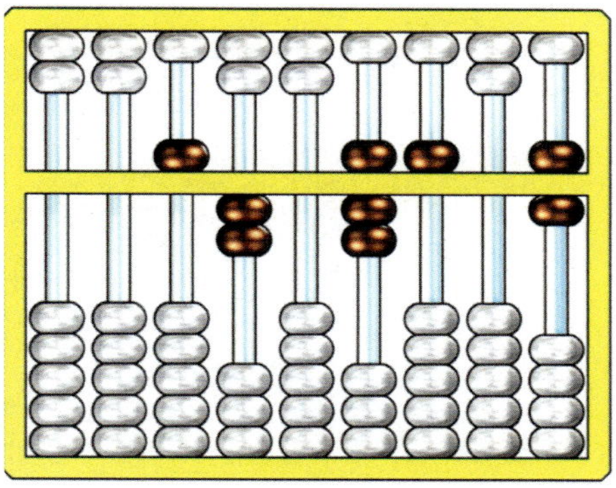

8. 4536 - 224 - 12 = **4300**

4536 - 224 = **4312**

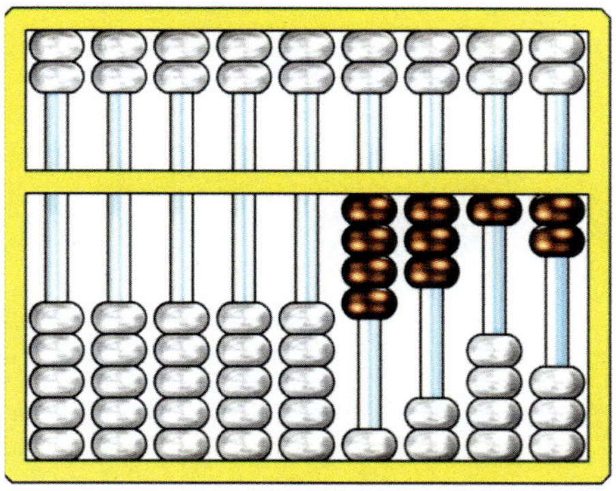

4312 - 12 = **4300**

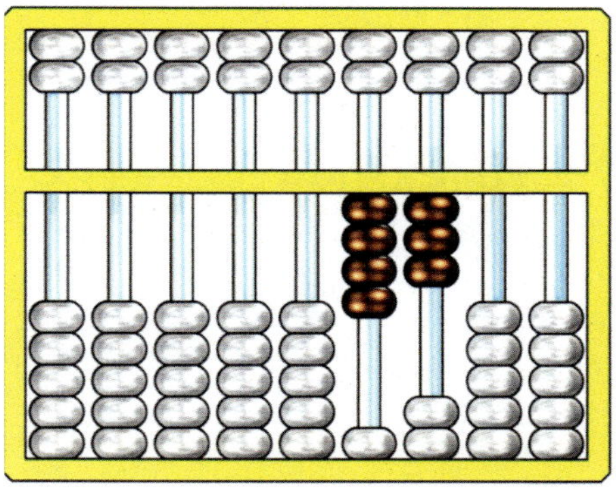

Answers to test 4 (on page 63)

9

440262 - 3201 - 435 - 6713 = **429913**

440262 - 3201 = **437061**

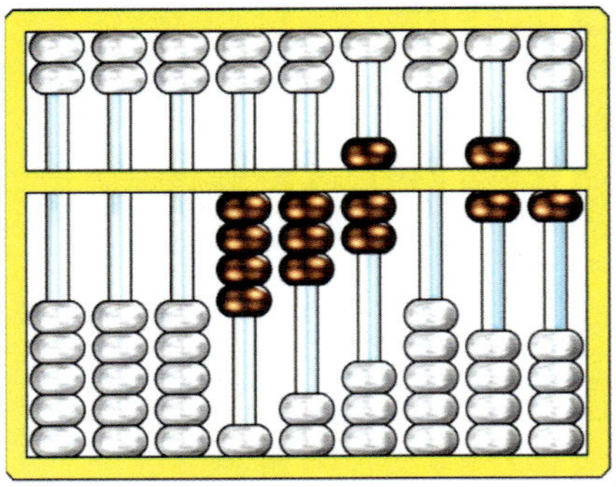

437061 - 435 = **436626**

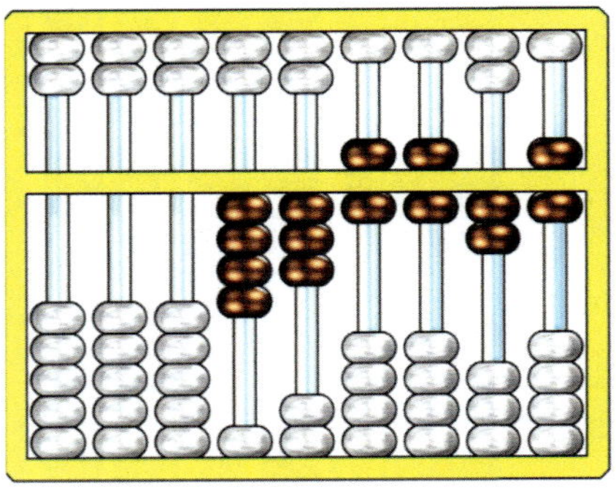

436626 - 6713 = **429913**

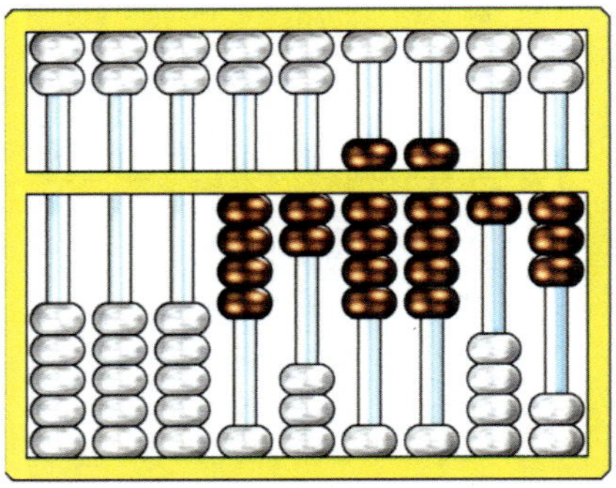

10

87654321 - 10000008 - 642107 = **77012206**

87654321 - 10000008 = **77654313**

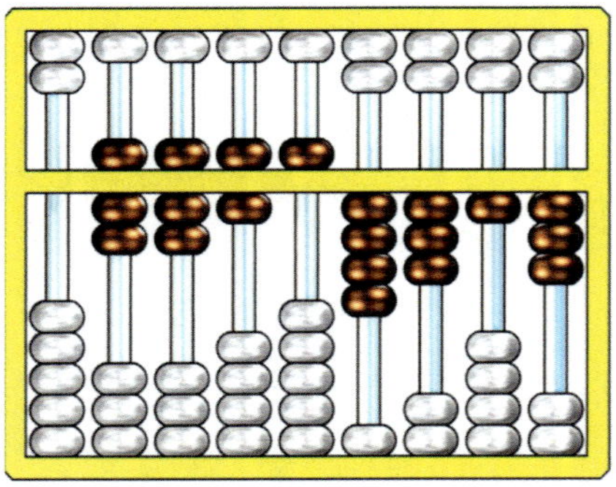

77654313 - 642107 = **77012206**

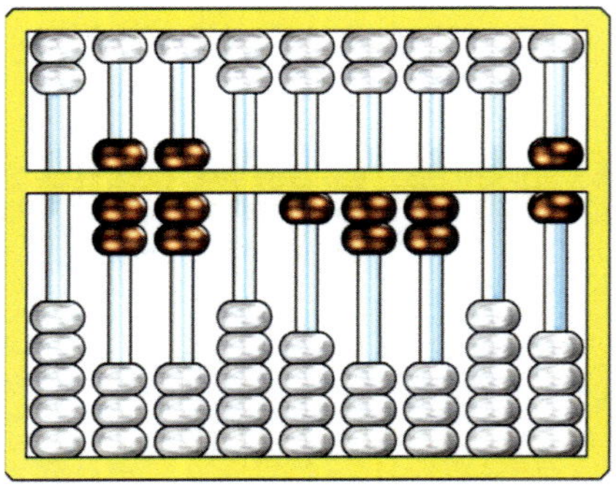

Subtraction of decimal numbers

Remember that a decimal number is a number that contains a decimal point, like 0.5 or 0.24

When we put a decimal number on the abacus, remember we place a dot on the beam to mark our ones column, so any number to the right of the dot (columns 1,2 & 3) will be our decimals.

Example: 4.623 - 1.21

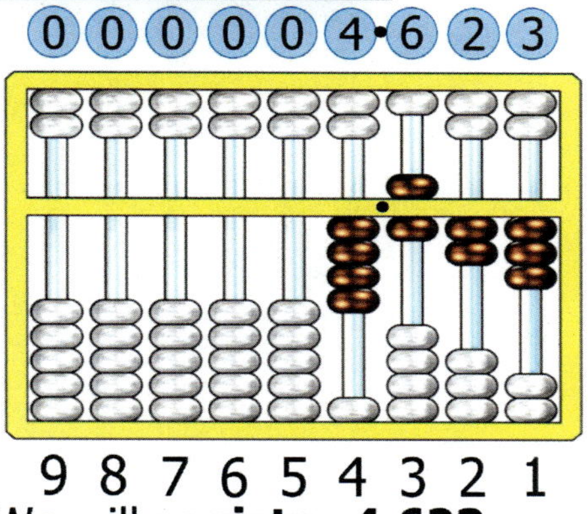

We will **register 4.623**

4	• Column 4, register 4 lower beads
0.6	• Column 3, register 1 upper and 1 lower bead
0.02	• Column 2, register 2 lower beads
0.003	• Column 1, register 3 lower beads

The abacus reads 4.623

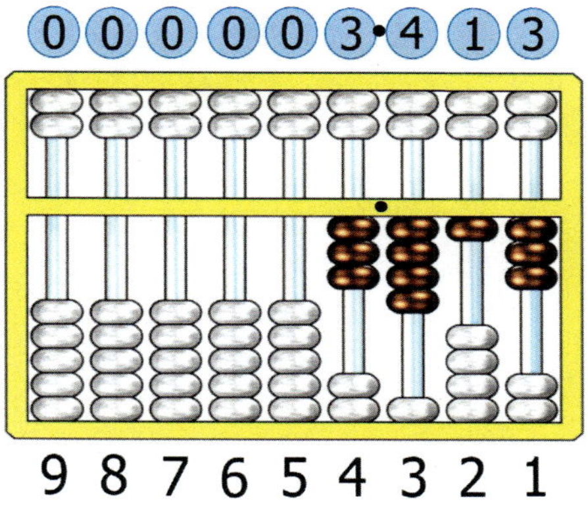

We will **subtract 1.21**

-1	• Column 4, unregister 1 lower bead
-0.2	• Column 3, unregister 1 upper bead and register 3 lower beads
-0.01	• Column 2, unregister 1 lower bead
	• Column 1, do nothing

The abacus result is 3.413

Example: 564.752 - 82.6

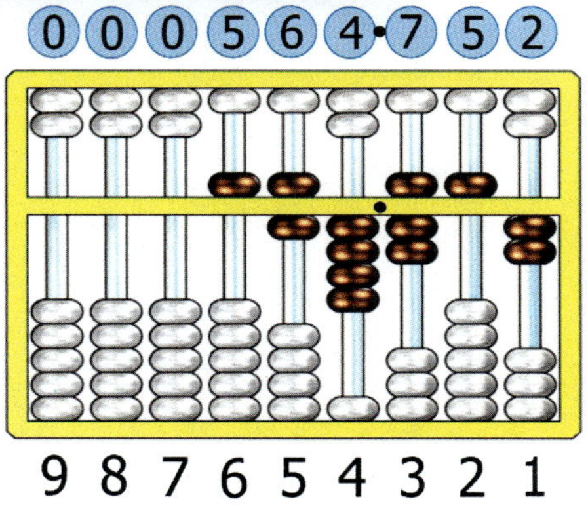

We will **register 564.752**

500	• Column 6, register 1 upper bead
60	• Column 5, register 1 upper and 1 lower bead
4	• Column 4, register 4 lower beads
0.7	• Column 3, register 1 upper and 2 lower beads
0.05	• Column 2, register 1 upper bead
0.002	• Column 1, register 2 lower beads

The abacus reads 564.752

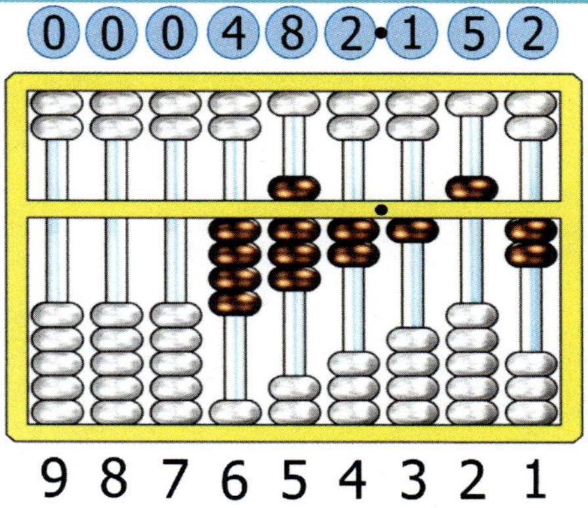

We will **subtract 82.6**

There are not enough beads in column 5, move to column 6, think **-8=-10+2**

-100	• Column 6, unregister 1 upper bead and register 4 lower beads
+20	• Column 5, register 2 lower beads
-2	• Column 4, unregister 2 lower beads
-0.6	• Column 3, unregister 1 upper and 1 lower bead
	• Columns 2 & 1, do nothing

The abacus result is 482.152

More examples of subtraction of decimal numbers

Example: 69.799 - 3.461

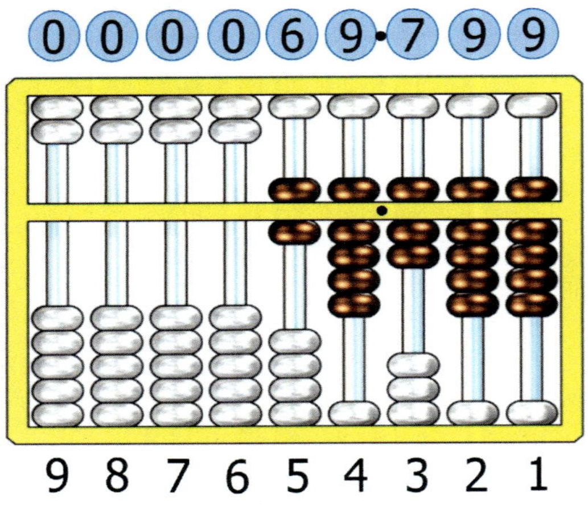

We will **register 69.799**

60	• Column 5, register 1 upper and 1 lower bead
9	• Column 4, register 1 upper and 4 lower beads
0.7	• Column 3, register 1 upper and 2 lower beads
0.09	• Column 2, register 1 upper and 4 lower beads
0.009	• Column 1, register 1 upper and 4 lower beads

The abacus reads 69.799

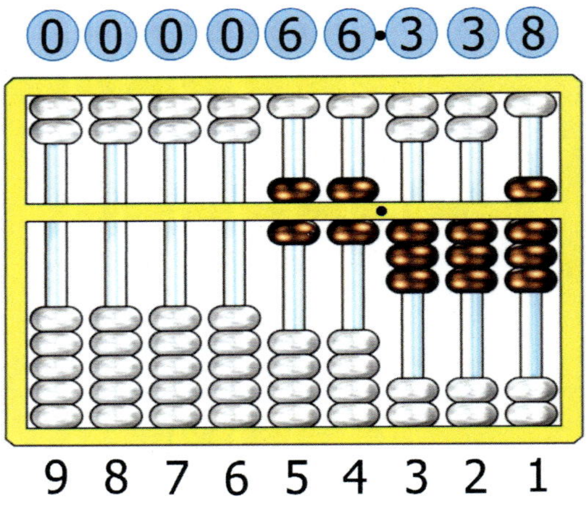

We will now **subtract 3.461**

-3	• Column 4, unregister 3 lower beads
-0.4	• Column 3, unregister 1 upper and register 1 lower bead
-0.06	• Column 2, unregister 1 upper and 1 lower bead
-0.001	• Column 1, unregister 1 lower bead

The abacus result is 66.338

TEST 5 - Subtraction of decimal numbers

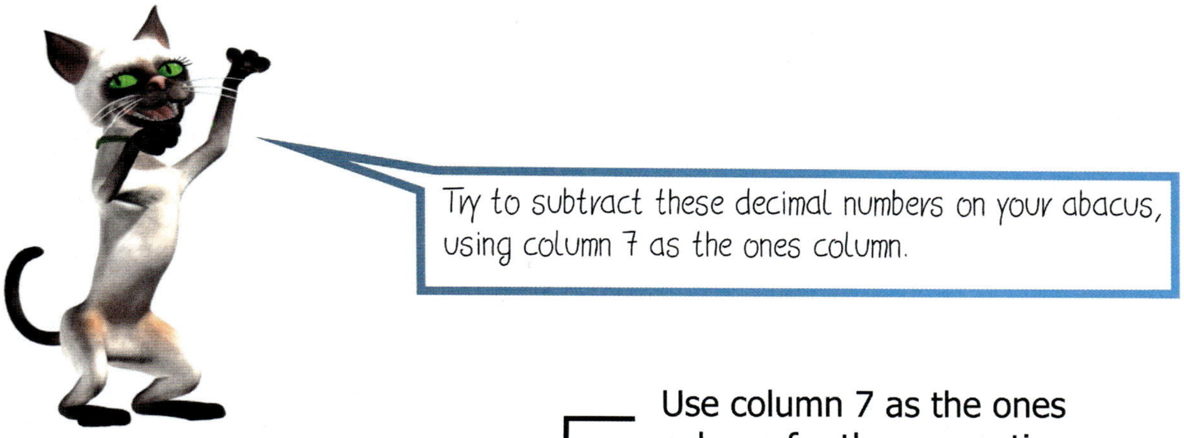

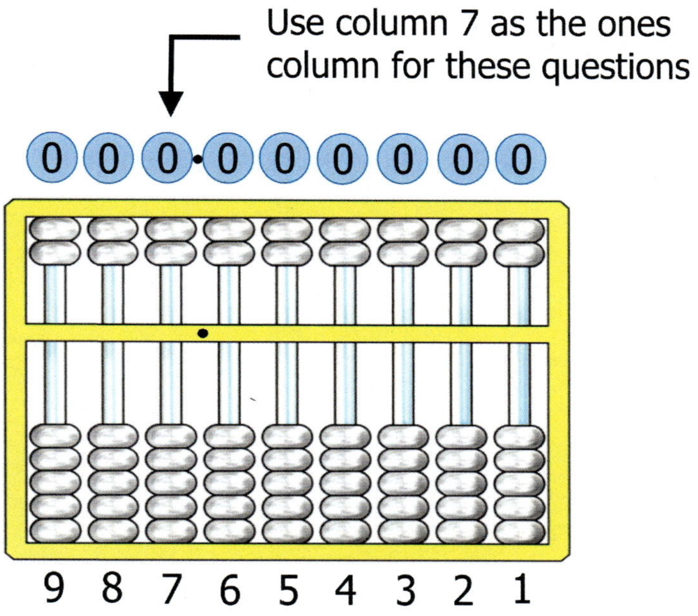

Answers are on page 70.

1. 1.601091 - 1.2119

2. 399.8898 - 110.8802

3. 6.6011 - 5.31321

4. 10.586 - 10.123

5. 999.998621 - 12.3443

Answers to test 5 (on page 69)

Here are the subtraction of decimal numbers answers.
If you got any wrong, just have another go.

1

1.601091 - 1.2119 = **0.389191**

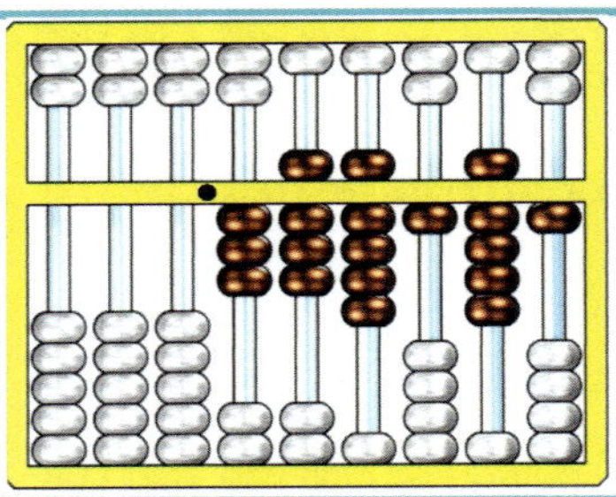

2

399.8898 - 110.8802 = **289.0096**

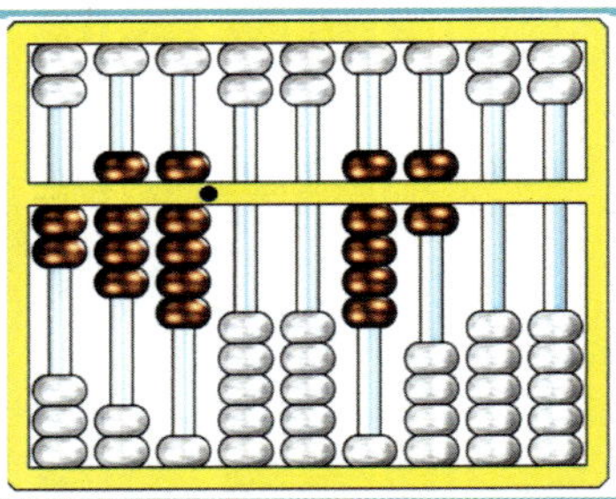

3

6.6011 - 5.31321 = **1.28789**

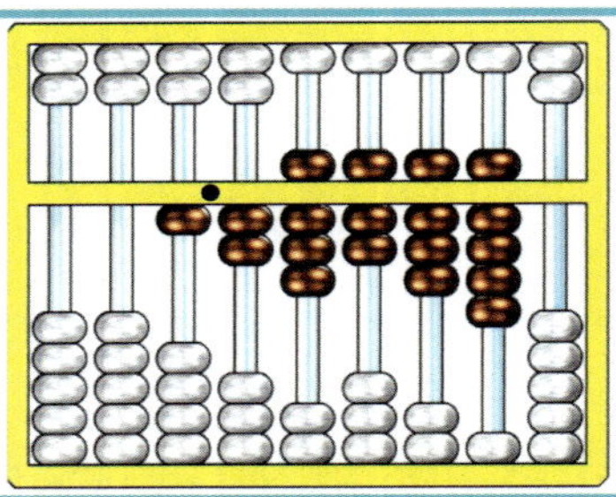

4

10.586 - 10.123 = **0.463**

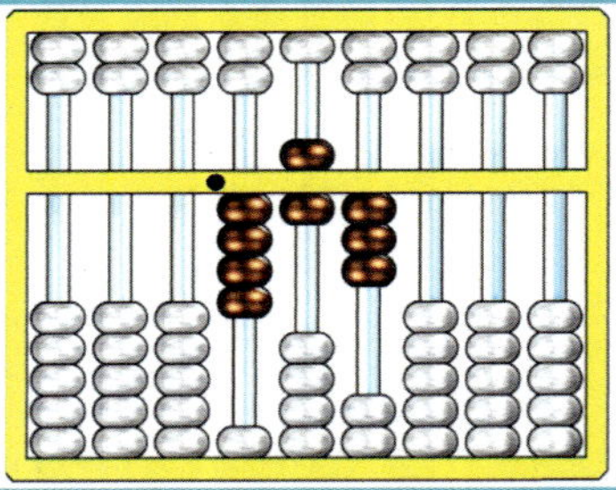

5

999.998621 - 12.3443 = **987.654321**

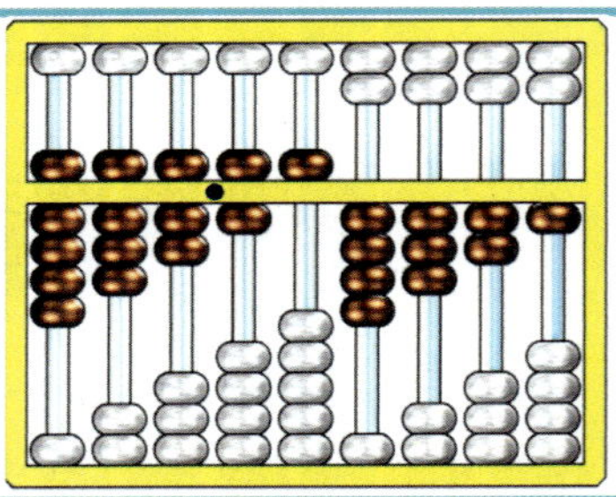

Well done!!!
You've finished the course.
I can use the abacus too!!!

MY NOTES

Made in the USA
Monee, IL
06 December 2021

84059204R00045